Pitman Research Notes in Mathematics Series

Submission of proposals for consideration
Suggestions for publication, in the form of outlines and representative samples, are invited by the Editorial Board for assessment. Intending authors should approach one of the main editors or another member of the Editorial Board, citing the relevant AMS subject classifications. Alternatively, outlines may be sent directly to the publisher's offices. Refereeing is by members of the board and other mathematical authorities in the topic concerned, throughout the world.

Preparation of accepted manuscripts
On acceptance of a proposal, the publisher will supply full instructions for the preparation of manuscripts in a form suitable for direct photo-lithographic reproduction. Specially printed grid sheets are provided and a contribution is offered by the publisher towards the cost of typing. Word processor output, subject to the publisher's approval, is also acceptable.

Illustrations should be prepared by the authors, ready for direct reproduction without further improvement. The use of hand-drawn symbols should be avoided wherever possible, in order to maintain maximum clarity of the text.

The publisher will be pleased to give any guidance necessary during the preparation of a typescript, and will be happy to answer any queries.

Important note
In order to avoid later retyping, intending authors are strongly urged not to begin final preparation of a typescript before receiving the publisher's guidelines and special paper. In this way it is hoped to preserve the uniform appearance of the series.

Longman Scientific & Technical
Longman House
Burnt Mill
Harlow, Essex, UK
(tel (0279) 26721)

Titles in this series

1 Improperly posed boundary value problems
A Carasso and A P Stone

2 Lie algebras generated by finite dimensional ideals
I N Stewart

3 Bifurcation problems in nonlinear elasticity
R W Dickey

4 Partial differential equations in the complex domain
D L Colton

5 Quasilinear hyperbolic systems and waves
A Jeffrey

6 Solution of boundary value problems by the method of integral operators
D L Colton

7 Taylor expansions and catastrophes
T Poston and I N Stewart

8 Function theoretic methods in differential equations
R P Gilbert and R J Weinacht

9 Differential topology with a view to applications
D R J Chillingworth

10 Characteristic classes of foliations
H V Pittie

11 Stochastic integration and generalized martingales
A U Kussmaul

12 Zeta-functions: An introduction to algebraic geometry
A D Thomas

13 Explicit *a priori* inequalities with applications to boundary value problems
V G Sigillito

14 Nonlinear diffusion
W E Fitzgibbon III and H F Walker

15 Unsolved problems concerning lattice points
J Hammer

16 Edge-colourings of graphs
S Fiorini and R J Wilson

17 Nonlinear analysis and mechanics: Heriot-Watt Symposium Volume I
R J Knops

18 Actions of fine abelian groups
C Kosniowski

19 Closed graph theorems and webbed spaces
M De Wilde

20 Singular perturbation techniques applied to integro-differential equations
H Grabmüller

21 Retarded functional differential equations: A global point of view
S E A Mohammed

22 Multiparameter spectral theory in Hilbert space
B D Sleeman

24 Mathematical modelling techniques
R Aris

25 Singular points of smooth mappings
C G Gibson

26 Nonlinear evolution equations solvable by the spectral transform
F Calogero

27 Nonlinear analysis and mechanics: Heriot-Watt Symposium Volume II
R J Knops

28 Constructive functional analysis
D S Bridges

29 Elongational flows: Aspects of the behaviour of model elasticoviscous fluids
C J S Petrie

30 Nonlinear analysis and mechanics: Heriot-Watt Symposium Volume III
R J Knops

31 Fractional calculus and integral transforms of generalized functions
A C McBride

32 Complex manifold techniques in theoretical physics
D E Lerner and P D Sommers

33 Hilbert's third problem: scissors congruence
C-H Sah

34 Graph theory and combinatorics
R J Wilson

35 The Tricomi equation with applications to the theory of plane transonic flow
A R Manwell

36 Abstract differential equations
S D Zaidman

37 Advances in twistor theory
L P Hughston and R S Ward

38 Operator theory and functional analysis
I Erdelyi

39 Nonlinear analysis and mechanics: Heriot-Watt Symposium Volume IV
R J Knops

40 Singular systems of differential equations
S L Campbell

41 N-dimensional crystallography
R L E Schwarzenberger

42 Nonlinear partial differential equations in physical problems
D Graffi

43 Shifts and periodicity for right invertible operators
D Przeworska-Rolewicz

44 Rings with chain conditions
A W Chatters and C R Hajarnavis

45 Moduli, deformations and classifications of compact complex manifolds
D Sundararaman

46 Nonlinear problems of analysis in geometry and mechanics
M Atteia, D Bancel and I Gumowski

47 Algorithmic methods in optimal control
W A Gruver and E Sachs

48 Abstract Cauchy problems and functional differential equations
F Kappel and W Schappacher

49 Sequence spaces
W H Ruckle

50 Recent contributions to nonlinear partial differential equations
H Berestycki and H Brezis

51 Subnormal operators
J B Conway

52 Wave propagation in viscoelastic media
F Mainardi
53 Nonlinear partial differential equations and their applications: Collège de France Seminar. Volume I
H Brezis and J L Lions
54 Geometry of Coxeter groups
H Hiller
55 Cusps of Gauss mappings
T Banchoff, T Gaffney and C McCrory
56 An approach to algebraic K-theory
A J Berrick
57 Convex analysis and optimization
J-P Aubin and R B Vintner
58 Convex analysis with applications in the differentiation of convex functions
J R Giles
59 Weak and variational methods for moving boundary problems
C M Elliott and J R Ockendon
60 Nonlinear partial differential equations and their applications: Collège de France Seminar. Volume II
H Brezis and J L Lions
61 Singular systems of differential equations II
S L Campbell
62 Rates of convergence in the central limit theorem
Peter Hall
63 Solution of differential equations by means of one-parameter groups
J M Hill
64 Hankel operators on Hilbert space
S C Power
65 Schrödinger-type operators with continuous spectra
M S P Eastham and H Kalf
66 Recent applications of generalized inverses
S L Campbell
67 Riesz and Fredholm theory in Banach algebra
B A Barnes, G J Murphy, M R F Smyth and T T West
68 Evolution equations and their applications
F Kappel and W Schappacher
69 Generalized solutions of Hamilton-Jacobi equations
P L Lions
70 Nonlinear partial differential equations and their applications: Collège de France Seminar. Volume III
H Brezis and J L Lions
71 Spectral theory and wave operators for the Schrödinger equation
A M Berthier
72 Approximation of Hilbert space operators I
D A Herrero
73 Vector valued Nevanlinna Theory
H J W Ziegler
74 Instability, nonexistence and weighted energy methods in fluid dynamics and related theories
B Straughan
75 Local bifurcation and symmetry
A Vanderbauwhede
76 Clifford analysis
F Brackx, R Delanghe and F Sommen
77 Nonlinear equivalence, reduction of PDEs to ODEs and fast convergent numerical methods
E E Rosinger
78 Free boundary problems, theory and applications. Volume I
A Fasano and M Primicerio
79 Free boundary problems, theory and applications. Volume II
A Fasano and M Primicerio
80 Symplectic geometry
A Crumeyrolle and J Grifone
81 An algorithmic analysis of a communication model with retransmission of flawed messages
D M Lucantoni
82 Geometric games and their applications
W H Ruckle
83 Additive groups of rings
S Feigelstock
84 Nonlinear partial differential equations and their applications: Collège de France Seminar. Volume IV
H Brezis and J L Lions
85 Multiplicative functionals on topological algebras
T Husain
86 Hamilton-Jacobi equations in Hilbert spaces
V Barbu and G Da Prato
87 Harmonic maps with symmetry, harmonic morphisms and deformations of metrics
P Baird
88 Similarity solutions of nonlinear partial differential equations
L Dresner
89 Contributions to nonlinear partial differential equations
C Bardos, A Damlamian, J I Díaz and J Hernández
90 Banach and Hilbert spaces of vector-valued functions
J Burbea and P Masani
91 Control and observation of neutral systems
D Salamon
92 Banach bundles, Banach modules and automorphisms of C*-algebras
M J Dupré and R M Gillette
93 Nonlinear partial differential equations and their applications: Collège de France Seminar. Volume V
H Brezis and J L Lions
94 Computer algebra in applied mathematics: an introduction to MACSYMA
R H Rand
95 Advances in nonlinear waves. Volume I
L Debnath
96 FC-groups
M J Tomkinson
97 Topics in relaxation and ellipsoidal methods
M Akgül
98 Analogue of the group algebra for topological semigroups
H Dzinotyiweyi
99 Stochastic functional differential equations
S E A Mohammed

100 Optimal control of variational inequalities
V Barbu
101 Partial differential equations and dynamical systems
W E Fitzgibbon III
102 Approximation of Hilbert space operators. Volume II
C Apostol, L A Fialkow, D A Herrero and D Voiculescu
103 Nondiscrete induction and iterative processes
V Ptak and F-A Potra
104 Analytic functions – growth aspects
O P Juneja and G P Kapoor
105 Theory of Tikhonov regularization for Fredholm equations of the first kind
C W Groetsch
106 Nonlinear partial differential equations and free boundaries. Volume I
J I Díaz
107 Tight and taut immersions of manifolds
T E Cecil and P J Ryan
108 A layering method for viscous, incompressible L_p flows occupying R^n
A Douglis and E B Fabes
109 Nonlinear partial differential equations and their applications: Collège de France Seminar. Volume VI
H Brezis and J L Lions
110 Finite generalized quadrangles
S E Payne and J A Thas
111 Advances in nonlinear waves. Volume II
L Debnath
112 Topics in several complex variables
E Ramírez de Arellano and D Sundararaman
113 Differential equations, flow invariance and applications
N H Pavel
114 Geometrical combinatorics
F C Holroyd and R J Wilson
115 Generators of strongly continuous semigroups
J A van Casteren
116 Growth of algebras and Gelfand–Kirillov dimension
G R Krause and T H Lenagan
117 Theory of bases and cones
P K Kamthan and M Gupta
118 Linear groups and permutations
A R Camina and E A Whelan
119 General Wiener–Hopf factorization methods
F-O Speck
120 Free boundary problems: applications and theory, Volume III
A Bossavit, A Damlamian and M Fremond
121 Free boundary problems: applications and theory, Volume IV
A Bossavit, A Damlamian and M Fremond
122 Nonlinear partial differential equations and their applications: Collège de France Seminar. Volume VII
H Brezis and J L Lions
123 Geometric methods in operator algebras
H Araki and E G Effros
124 Infinite dimensional analysis–stochastic processes
S Albeverio
125 Ennio de Giorgi Colloquium
P Krée
126 Almost-periodic functions in abstract spaces
S Zaidman
127 Nonlinear variational problems
A Marino, L Modica, S Spagnolo and M Degiovanni
128 Second-order systems of partial differential equations in the plane
L K Hua, W Lin and C-Q Wu
129 Asymptotics of high-order ordinary differential equations
R B Paris and A D Wood
130 Stochastic differential equations
R Wu
131 Differential geometry
L A Cordero
132 Nonlinear differential equations
J K Hale and P Martinez-Amores
133 Approximation theory and applications
S P Singh
134 Near-rings and their links with groups
J D P Meldrum
135 Estimating eigenvalues with *a posteriori/a priori* inequalities
J R Kuttler and V G Sigillito
136 Regular semigroups as extensions
F J Pastijn and M Petrich
137 Representations of rank one Lie groups
D H Collingwood
138 Fractional calculus
G F Roach and A C McBride
139 Hamilton's principle in continuum mechanics
A Bedford
140 Numerical analysis
D F Griffiths and G A Watson
141 Semigroups, theory and applications. Volume I
H Brezis, M G Crandall and F Kappel
142 Distribution theorems of L-functions
D Joyner
143 Recent developments in structured continua
D De Kee and P Kaloni
144 Functional analysis and two-point differential operators
J Locker
145 Numerical methods for partial differential equations
S I Hariharan and T H Moulden
146 Completely bounded maps and dilations
V I Paulsen
147 Harmonic analysis on the Heisenberg nilpotent Lie group
W Schempp
148 Contributions to modern calculus of variations
L Cesari
149 Nonlinear parabolic equations: qualitative properties of solutions
L Boccardo and A Tesei
150 From local times to global geometry, control and physics
K D Elworthy

151 A stochastic maximum principle for optimal control of diffusions
U G Haussmann
152 Semigroups, theory and applications. Volume II
H Brezis, M G Crandall and F Kappel
153 A general theory of integration in function spaces
P Muldowney
154 Oakland Conference on partial differential equations and applied mathematics
L R Bragg and J W Dettman
155 Contributions to nonlinear partial differential equations. Volume II
J I Díaz and P L Lions
156 Semigroups of linear operators: an introduction
A C McBride
157 Ordinary and partial differential equations
B D Sleeman and R J Jarvis
158 Hyperbolic equations
F Colombini and M K V Murthy
159 Linear topologies on a ring: an overview
J S Golan
160 Dynamical systems and bifurcation theory
M I Camacho, M J Pacifico and F Takens
161 Branched coverings and algebraic functions
M Namba
162 Perturbation bounds for matrix eigenvalues
R Bhatia
163 Defect minimization in operator equations: theory and applications
R Reemtsen
164 Multidimensional Brownian excursions and potential theory
K Burdzy
165 Viscosity solutions and optimal control
R J Elliott
166 Nonlinear partial differential equations and their applications. Collège de France Seminar. Volume VIII
H Brezis and J L Lions
167 Theory and applications of inverse problems
H Haario
168 Energy stability and convection
G P Galdi and B Straughan
169 Additive groups of rings. Volume II
S Feigelstock
170 Numerical analysis 1987
D F Griffiths and G A Watson
171 Surveys of some recent results in operator theory. Volume I
J B Conway and B B Morrel
172 Amenable Banach algebras
J-P Pier
173 Pseudo-orbits of contact forms
A Bahri
174 Poisson algebras and Poisson manifolds
K H Bhaskara and K Viswanath
175 Maximum principles and eigenvalue problems in partial differential equations
P W Schaefer
176 Mathematical analysis of nonlinear, dynamic processes
K U Grusa
177 Cordes' two-parameter spectral representation theory
D F McGhee and R H Picard
178 Equivariant K-theory for proper actions
N C Phillips
179 Elliptic operators, topology and asymptotic methods
J Roe
180 Nonlinear evolution equations
J K Engelbrecht, V E Fridman and E N Pelinovski
181 Nonlinear partial differential equations and their applications. Collège de France Seminar. Volume IX
H Brezis and J L Lions
182 Critical points at infinity in some variational problems
A Bahri
183 Recent developments in hyperbolic equations
L Cattabriga, F Colombini, M K V Murthy and S Spagnolo
184 Optimization and identification of systems governed by evolution equations on Banach space
N U Ahmed
185 Free boundary problems: theory and applications. Volume I
K H Hoffmann and J Sprekels
186 Free boundary problems: theory and applications. Volume II
K H Hoffmann and J Sprekels
187 An introduction to intersection homology theory
F Kirwan
188 Derivatives, nuclei and dimensions on the frame of torsion theories
J S Golan and H Simmons
189 Theory of reproducing kernels and its applications
S Saitoh
190 Volterra integrodifferential equations in Banach spaces and applications
G Da Prato and M Iannelli
191 Nest algebras
K R Davidson
192 Surveys of some recent results in operator theory. Volume II
J B Conway and B B Morrel
193 Nonlinear variational problems. Volume II
A Marino and M K Murthy
194 Stochastic processes with multidimensional parameter
M E Dozzi
195 Prestressed bodies
D Iesan
196 Hilbert space approach to some classical transforms
R H Picard
197 Stochastic calculus in application
J R Norris

Hilbert space approach to some classical transforms

Rainer Picard

University of Wisconsin-Milwaukee

Hilbert space approach to some classical transforms

Longman Scientific & Technical

Copublished in the United States with
John Wiley & Sons, Inc., New York

Longman Scientific & Technical,
Longman Group UK Limited,
Longman House, Burnt Mill, Harlow
Essex CM20 2JE, England
and Associated Companies throughout the world.

Copublished in the United States with
John Wiley & Sons, Inc., 605 Third Avenue, New York, NY 10158

First published 1989

AMS Subject Classifications: (Main) 44-02, 46-02, 42-02
(Subsidiary) 34-02, 33-02

ISSN 0269-3674

British Library Cataloguing in Publication Data
Picard, Ranier
A Hilbert space approach to some classical transforms.
1. Hilbert spaces. Self-adjoint operators.
Spectral representation.
I. Title.
515.7'33

ISBN 0-582-03189-3

Library of Congress Cataloging-in-Publication Data
Picard, R. H. (Rainer H.)
A. Hilbert space approach to some classical transforms.
(196)
Lecture notes based on a course held in the winter semester 1983/84 at the University of Bonn.
Includes index.
I. Integral transforms. 2. Spectral theory (Mathematics)
3. Hilbert space. I. Title. II. Series: Pitman
research notes in mathematics series; 196
QA432.P45 1988 515.7'23 88-8464

ISBN 0-470-21274-8

Printed and bound in Great Britain
by Biddles Ltd, Guildford and King's Lynn

Contents

Preface

The following lecture notes are based on a course held in the winter semester 1983/84 at the University of Bonn, Fed. Rep. of Germany. They are an extented version of a previous manuscript with the title 'On some special functions of mathematical physics and their relation to spectral theory', [27]. The response to this preliminary version encouraged the author to prepare the present expanded version. One idea the following is based on is the observation that some of the well-known integral transforms are closely connected with the spectral theory of a differential operator. It turns out that these integral transforms can in fact be "guessed" from the spectral resolution of an appropriately defined selfadjoint realization of this differential operator without using the machinery of Titchmarch-Weyl-Kodaira theory. This idea has been employed systematically in a paper by R. Leis from 1979, [20] (compare also [5], pp. 31-36). In the light of this approach the discussion of many integral transforms may be considered as immediate application of general spectral theory of selfadjoint operators. As is more commonly known many classical expansion theorems are results about eigenfunction expansion with respect to certain selfadjoint differential operators. A few examples of this type are included in the lecture notes. The main emphasis is however on reconstructing special integral transforms like e.g. the Laplace transform and the Hankel transform as spectral theorems.

Integral transforms (and eigensolution expansions) are considered in the following as constructive versions of an abstract spectral representation (i.e. a unitary equivalence between the selfadjoint differential operator in question and the operator of multiplication by the argument), the existence of which is assured by general spectral theory of selfadjoint operators. The fact that a spectral representation is in particular an isometry is displayed in what is known as Parseval's equality and therefore becomes a central concern. Parseval's equality provides the means to extend the representation of the spectral family initially obtained only for a dense subset to the whole underlying Hilbert space. The question of completeness arises naturally as the result establishing that the representation found is in fact not only an isometry but also onto (and thus unitary). The concept of unitarity provides us with natural characterizations of domains and ranges of the respective integral transforms.

Generally speaking an integral transform is a bounded mapping from a function space X onto a function space Y that together with its inverse has the form of an integral operator. The approach presented here is apparently restricted to a class of unitary integral transforms, namely those obtainable from the spectral representation of selfadjoint realizations of certain ordinary differential expressions. It is therefore only natural to limit our investigation to the Hilbert space case; for distribution theoretical approaches to the topic see e.g. [4], [41], [42] (compare also [31] for a Banach space approach).

Spectral representations provide the means to establish a function calculus. It is interesting to note that some integral transforms that do not seem to be spectral representations for a selfadjoint differential operator turn in fact out to be functions of such an operator. This observation seems to be known (see [5], p. 35-36, for the operator $i\frac{d}{dx}$ and convolutions) but has —— to the author's knowledge —— not been employed consistently as a theoretical framework to discuss special integral transforms.

It also turns out that some of the classicaly known relationships between different integral transforms can be explained by the fact that the corresponding differential operators are unitarily eqivalent. We shall see that all integral transforms related to first order differential expressions are in fact simple modifications of the Fourier transform. Therefore the Fourier transform becomes a central topic in this context. It is an interesting fact that despite the self-imposed limitation to a Hilbert space approach there is a natural access to extensions of the Fourier transform to certain distribution spaces which we shall call tempered Sobolev spaces. This approach has been inspired by the concept of tempered derivatives of generalized functions described in the book by P. Antosik, J. Mikusinski and R. Sikorski , [4]. The possibility to define tempered distributions in terms of tempered derivatives of functions is the basis of our construction of tempered Sobolev spaces.

In our discussion of special integral transforms completeness is intended only in the sense that at least the examples listed in the book of W. Magnus, F. Oberhettinger and R.P. Soni, [23], p. 396-398, are covered.

The discussion if there are other classes of functions to which these integral transforms can be applied (extensions of the domain) and other values of parameters for which the transformed function is still meaningful (extensions of the range) are classified as specific special function topics and are left out of consideration. For questions of this type the interested reader may be referred to the quoted special function literature as well as the literature quoted there (e.g. [19], [23], [25], [32], [34], [35], [41] and [42]).

These lecture notes are certainly not intended as a replacement of investigations of special function of mathematical physics but rather an approach of some aspects of this area from a certain point of view; in fact, the study of related special functions as solutions of certain ordinary differential equations is prior to the considerations presented here and we have chosen to quote corresponding results rather than to derive them. In particular the knowledge of the asymptotic behaviour of some special functions is of importance in the discussion of spectral properties of the underlying selfadjoint differential operator. The necessary results are quoted mainly from [23]. Some of the examples presented in these notes are dealt with at other places but typically using the Titchmarch-Weyl-Kodaira approach. The theory and examples may be found in several textbooks like e.g. [3] appendix 2, [11] chapter XIII, [15], [35], [37] chapter 8.4, [40] chapter 5.

The emphasis of these lecture notes — despite the considerable detail in parts of the presentation — is on ideas rather than techniques. This means in particular that many assumptions are chosen for convenience rather than to obtain the farthest generality. The reader will find possibilities of generalizations indicated but usually not executed. We like to look at our particular approach as one way to establish the means to bring order into the zoo of special functions. Since we are merely applying abstract spectral theory to special examples, these notes may also be regarded as a source of examples and exercises in functional analysis.

This monograph is divided into three main parts. The first part deals with the differential expression $D = \frac{1}{2\pi i}\frac{d}{dx}$ and related expansion theorems and integral transforms. In particular, we are led to the discussion of the Fourier transform and its 'relatives'. Because of its importance in this context the discussion of the Fourier transform takes up a major part. We also include the higher dimensional Fourier transform as a multiparameter spectral reperesentation formula. It provides us with a model to construct higher-dimensional transforms from one-dimensional ones.

The second part contains examples for second order differential expressions (particular Sturm-Liouville operators). After an initial brief discussion of classical polynomials we turn our attention to integral transforms again. It turns out that in order to reconstruct well-known integral transforms it suffices to consider abstract Neumann type problems. These are problems of the type where the selfadjoint realization A of the second order differential expression is of the form

$$A = C\,C^{*},$$

here C is a (closed) first order differential operator with smooth functions having compact support in the underlying interval as a dense set of the domain (i.e. a core). The operator C^{*} denotes its adjoint with respect to suitably chosen weighted L_2-spaces as initial and terminal space for C.

For illustrative purposes a fourth order example of a similar structure that goes back to a paper by Mercer, [24], and has been discussed previously in a similar setting in [20] is included.

The appendix gives a survey of material from functional analysis that is being applied in or useful for understanding the other two parts. It is not intended as an introduction to functional analysis but rather a compendium roughly indicating the area of functional analysis of which at least superficial acquaintance is required. It has been attempted to provide a complete reference source for the functional analytical concepts used in the rest of the presentation.

Finally I would like to express my gratitude to Professors Gary Roach and Erhard Meister for their encouragement and support, and in particular Professor Rolf Leis, also for his helpful comments during the preparation of the final draft.

University of Wisconsin-Milwaukee Rainer Picard

Part 1: Spectral representation results related to the differential expression $D = \dfrac{1}{2\pi i}\dfrac{d}{dx}$

1.1 The Hilbert space approach for the differential operator D

The first example we want to consider is the simplest differential expression one can think of, namely $D = \frac{1}{2\pi i}\frac{d}{dx}$. We start our approach by defining the following operator

$$\begin{aligned} d :\ & C_\infty(I) \subset L_2(I) \longrightarrow L_2(I) , \\ & u \longrightarrow D\,u , \end{aligned} \tag{1.1}$$

$C_\infty(I)$ denotes the set of infinitely often differentiable, complex-valued functions Φ defined on an open interval I of the real line $\mathbb{R}$ having compact support ($\operatorname{supp}\Phi := \overline{\{x \in I \mid \Phi(x) \neq 0\}}$) in I . Here $L_2(I)$ denotes —— as usual —— the Hilbert space of measurable, square integrable functions defined on the interval I (with inner product:

$$(u,v) = \int_I \overline{u(x)}\, v(x)\, dx \quad \text{for } u,v \in L_2(I)),$$

and may be defined as the closure of $C_\infty(I)$ w.r.t. $\|\bullet\| = (\bullet,\bullet)^{1/2}$. Thus $C_\infty(I)$ is dense in $L_2(I)$ and the operator d obviously is densely defined and symmetric.
Indeed we have

$$\begin{aligned} (D\,u,v) &= \int_I \overline{D\,u(x)} \bullet v(x)\, dx \\ &= \int_I \frac{1}{2\pi}\, i\, \frac{d}{dx}\overline{u(x)} \bullet v(x)\, dx \\ &= \int_I \overline{u(x)} \bullet \frac{1}{2\pi i}\frac{d}{dx} v(x)\, dx \\ &= (u,\, D\,v) \end{aligned} \tag{1.2}$$

for all $u,v \in C_\infty(I)$.

In order to determine the deficiency indices we need the following simple lemma $(\Phi' \equiv \frac{d}{dx}\Phi(x))$.

<u>Lemma 1.1</u>: *Let* $f \in L_2(I)$ *for some interval* $I \subset \mathbb{R}$ *and*

$$(\Phi', f) = 0 \qquad \textit{for all} \quad \Phi \in C_\infty(I) , \tag{1.3}$$

then we have

$$f = \text{const} \quad \textit{a.e. in } I .$$

<u>Remark 1.1</u>: *The lemma can be extended to general distributions on the real line by similar arguments. Here we only make use of the version above. The result is infact a regularity theorem and shall be employed to show that there is no other solution of* (1.3) *then the one classically expected.*

<u>Proof of Lemma 1.1:</u>

For $\varphi \in C_\infty(I)$ and $x \in I$ let

$$\Phi(x) := \int_{I\cap(-\infty,x]} \varphi(t)\, dt \;-\; \int_{I\cap(-\infty,x]} \alpha(t)\, dt \int_I \varphi(t)\, dt ,$$

$\alpha \in C_\infty(I)$ choosen such that

$$\int_I \alpha(t)\, dt = 1 .$$

Then we have $\Phi \in C_\infty(I)$ and from (1.3) we get

$$(\varphi - \alpha \int_I \varphi(t)\, dt , f) = 0 , \tag{1.4}$$

for all $\varphi \in \overset{o}{C}_\infty(I)$. Note that Φ has been obtained by modifying a solution of $\Phi' = \varphi$ to have $\Phi \in C_\infty(I)$. This idea will be used at several occasions in the following.As a consequence of (1.4) we have for all $\varphi \in C_\infty(I)$

$$(\varphi, f) = \int_I \overline{\varphi(t)}\, (\alpha, f)\, dt$$

$$= (\varphi, (\alpha, f)) .$$

In other words

$f = (\alpha, f)$ a.e. ,

The number (α, f) is the constant mentioned in the lemma. □

In order to determine the deficiency indices we have to calculate the number of solutions in the null space $N(d^* \pm i)$ of $d^* \pm i$:

$$f \in N(d^* \pm i) \text{ iff } f \in L_2(I) \text{ and } ((d \mp i)\Phi, f) = 0 \text{ for all } \Phi \in C_\infty(I).$$

For such an $f \in L_2(I)$ we have

$$(\Phi', f) \pm 2\pi (\Phi, f) = 0 \text{ for all } \Phi \in C_\infty(I) .$$

We shall employ the following notation for pre-multiplying a function g by a function f in order to avoid confusion between free and bound variables

$$(f(m)\ g)(x) = f(x)\ g(x),\ x \in I,$$

where f,g are functions defined on I. For convenience we shall also identify $f = f(m)\ 1 = f(m)$. With this notation we replace Φ by $e^{\mp 2\pi m}\ \Psi$ then

$$(e^{\mp 2\pi m}\ \Psi', f) = 0 \text{ for all } \Psi \in C_\infty(I) ,$$

or by Lemma 1.1

$$e^{\mp 2\pi m}\ f = \text{const} ,$$

i.e.

$$f = C\ e^{\pm 2\pi m} \quad \text{for some } C \in \mathbb{R} .$$

In the analysis of d we have to distinguish 3 cases

1) I bounded,
2) I unbounded and the complement I' of I unbounded,
3) $I = \mathbb{R}$.

Case 1: $N(d^* \pm i) = \text{span}\{e^{\pm 2\pi m}\}$, thus

$$\alpha_\pm(d) := \dim N(d^* \pm i) = 1 .$$

Case 2: The interval I has the form (t_0, ∞) or $(-\infty, t_0)$ for some $t_0 \in \mathbb{R}$,

$$N(d^* + i) = \begin{cases} \text{span}\{e^{2\pi m}\} & \text{if I is bounded from above} \\ \{0\} & \text{if I is bounded from below,} \end{cases}$$

$$N(d^* - i) = \begin{cases} \{0\} & \text{if I is bounded from above} \\ \text{span}\{e^{-2\pi m}\} & \text{if I is bounded from below.} \end{cases}$$

Thus the deficiency indices $(\alpha_+|\alpha_-)$ are either $(1|0)$ or $(0|1)$ respectively. In any case, the closure of d is a maximally symmetric operator and does not allow for a selfadjoint extension.

Case 3: Here we have

$$N(d^* \pm i) = \{0\} ,$$

and so d is essentially selfadjoint in this case.

We shall deal with case 1) and case 3) in the following two sub-paragraphs.

<u>1.1.1 Selfadjoint realizations of the operator D in the case of a bounded interval. The Fourier series.</u>

Let d denote the closure $\bar{d}$ of d, then $D(d)$ may be characterized as the completion of $D(d) = C_\infty(I)$ w.r.t. the graph norm $\|\bullet\|_d$, where

$$\|u\|_d := \sqrt{\|u\|^2 + 4\pi^2\|d\,u\|^2} \quad \text{for} \quad u \in D(d), \ (\|u\|^2 := (u,u)).$$

We shall now give another description for $D(d)$ in "weak" terms. The following estimate is essential (compare [1], p. 32).

<u>Lemma 1.2</u> (Sobolev's inequality): *For all* $\Phi \in C_\infty((a,b))$ *the estimate*

$$|\Phi(x) - \Phi(y)| \leq |x-y|^{1/2} \left(\int_x^y \left|\frac{d}{dt}\Phi(t)\right|^2 dt\right)^{1/2}, \ b > y \geq x > a, \qquad (1.4)$$

is valid.

<u>Proof:</u> Estimate (1.4) is an immediate consequence of

$$\Phi(y) - \Phi(x) = \int_x^y \frac{d}{dt}\Phi(t)\,dt$$

and Cauchy-Schwarz's inequality.

□

$D(d)$ is a Hilbert space that will be denoted by $H_1(I)$. The norm in $H_1(I)$ is $\|\bullet\|_1 := \|\bullet\|_d$; or if the domain I is not obvious we shall write $\|\bullet\|_{1,I}$. From (1.4) we easily obtain that

$$H_1(I) \equiv D(d) \ \text{is boundedly imbedded in} \ C_{1/2}(I) . \qquad (1.5)$$

$C_{1/2}(I)$ is the space of continuous functions Φ on I with

$$|\Phi|_{1/2} := \sup_{x\in I} |\Phi(x)| + \sup_{\substack{x,y\in I\\ x\neq y}} \frac{|\Phi(x) - \Phi(y)|}{|x - y|^{1/2}} < \infty ,$$

a Banach space w.r.t. $|\cdot|_{1/2}$. From (1.4) we first have

$$|\Phi(x)| \leq L^{1/2} \, \| \frac{d}{dx} \Phi \|, \tag{1.6}$$

where $L := \sup_{x,y\in I} |x - y|$ is the length of the interval I. Consequently,

$$| \Phi |_{1/2} \leq (1 + L^{1/2}) \, \| \frac{d}{dx} \Phi \| \leq (1 + L^{1/2}) \, \| \Phi \|_1 \tag{1.7}$$

for all $\Phi \in C_\infty(I)$. By a closure argument we have indeed that $\Phi \in H_1(I)$ coincides a.e with a $C_{1/2}(I)$-function and estimate (1.7) is valid, i.e. (1.5). Furthermore, by (1.4) the following boundary condition is satisfied:

$$\lim_{x\to\partial I} (\operatorname{dist}(x,\partial I))^{-1/2} \Phi(x) = 0 \quad \text{for } \Phi \in H_1(I) . \tag{1.8}$$

Remark 1.2: *Note that by* (1.4)

$$(\operatorname{dist}(m,\partial I))^{-1/2} \Phi \textit{ is continuous for } \Phi \in D(d).$$

To be an element of $H_1(I)$ apparently involves a generalized boundary condition. This statement is in a certain way confirmed by the following density result.

Lemma 1.3: *The set of all functions in $C_\infty(\bar{I})$ having boundary values equal to zero is dense in $H_1(I)$.*

Proof: Let

$$\alpha(x) = C_0 \exp((x^2-1)^{-1}) \text{ for } x \in (-1, 1),$$

and define α to be identically zero elsewhere. Assume that C_0 is such that

$$\int_{\mathbb{R}} \alpha(t) \, dt = 1.$$

Apparently we have that $\alpha \in C_\infty((-2, 2))$. Define

$$\rho_\varepsilon(x) = \int_{(-\infty,x]} \alpha((t-a-2\varepsilon)/\varepsilon)\ \varepsilon^{-1}\ dt - \int_{(-\infty,x]} \alpha((t-b+2\varepsilon)/\varepsilon)\ \varepsilon^{-1}\ dt,$$

where $a,b \in \mathbb{R}$ are the boundary points of I, then

$$\rho_\varepsilon \in C_\infty(I) \text{ for small } \varepsilon > 0,$$

and

$$\rho_\varepsilon \longrightarrow 1 \text{ pointwise in I as } \varepsilon \longrightarrow 0.$$

Thus, let Ψ be any element in $C_\infty(\bar{I})$, then

$$\rho_\varepsilon\ \Psi \in C_\infty(I),\ \rho_\varepsilon\ \Psi \longrightarrow \Psi \text{ in } L_2(I) \text{ as } \varepsilon \longrightarrow 0.$$

$$\rho_\varepsilon\ \Psi' \longrightarrow \Psi' \text{ in } L_2(I) \text{ as } \varepsilon \longrightarrow 0.$$

Moreover, we have

$$\rho_\varepsilon'\ \Psi \longrightarrow 0 \text{ in } L_2(I) \text{ as } \varepsilon \longrightarrow 0.$$

This can be seen from the following calculation.

$$\|\rho_\varepsilon'\ \Psi\|^2 = \int_I |\rho_\varepsilon'(t)|^2\ |\Psi(t)|^2\ dt$$

$$= \int_{(a,a+2\varepsilon)} |\rho_\varepsilon'(t)|^2\ |\Psi(t)|^2\ dt + \int_{(b-2\varepsilon,b)} |\rho_\varepsilon'(t)|^2\ |\Psi(t)|^2\ dt$$

$$= \int_{(-1,\ 1)} |\alpha(s)|^2\ |\Psi(\varepsilon s + a + 2\varepsilon)|^2\ \varepsilon^{-1}\ ds + $$

$$+ \int_{(-1,\ 1)} |\alpha(s)|^2\ |\Psi(\varepsilon s + b - 2\varepsilon)|^2\ \varepsilon^{-1}\ ds.$$

The last two integrals can be estimated by

$$\varepsilon\ 2\ C_0 \max_{s\in I} |\Psi'(s)|^2.$$

This shows that

$$(\rho_\varepsilon\Psi)' = \rho_\varepsilon\Psi' + \rho_\varepsilon'\Psi \longrightarrow 0 \text{ as } \varepsilon \longrightarrow 0.$$

□

We observe that $D(d^*)$ is a Hilbert space w.r.t. the corresponding graph norm $\|\cdot\|_{d^*}$. We introduce a separate notation

$$W_1(I) := D(d^*) .$$

Extending the meaning of derivative to elements of $W_1(I)$ by setting

$$\frac{d}{dt}\Phi(t) := 2\pi i\, d^*\Phi(t) \text{ for } \Phi \in D(d^*) ,$$

we have

$$W_1(I) = \{\Phi \in L_2(I) \mid \Phi' \in L_2(I)\} .$$

This definition makes sense in the case $I = \mathbb{R}$ as well and will be used later. The norm will be denoted by $\|\cdot\|_1$ and we have by definition

$$\|\Phi\|_1^2 = \|\Phi\|^2 + \|\Phi'\|^2 = \|\Phi\|^2 + 4\pi^2\|d^*\Phi\|^2 = \|\Phi\|^2_{d^*} .$$

Since d^* is a extension of d (i.e. $d \subset d^*$; d is symmetric) we have that $H_1(I)$ may be regarded as a subspace of the Hilbert space $W_1(I)$. Obviously, we have

$$H_1(I) \subset \overline{C_\infty(I) \cap W_1(I)} \subset W_1(I) .$$

The boundary condition (1.8) and Lemma 1.3 indicate that $H_1(I)$ might be described as the set of those elements in $W_1(I)$ satisfying the boundary condition of vanishing boundary data.

In order to make this precise we need a "trace theorem". As a first step in this direction we observe that for the Hilbert space

$$H_1(I) := \overline{C_\infty(I) \cap W_1(I)}^{\|\cdot\|_1},$$

i.e. the closure in $W_1(I)$ of the set $C_\infty(I) \cap W_1(I)$, we have (compare e.g. [16], [1], p. 11, [21], p. 10):

<u>Lemma 1.4</u>: $$H_1(I) = W_1(I) .$$

<u>Proof</u>: We have to show that $C_\infty(I) \cap W_1(I)$ is dense in $W_1(I)$. Let $\Phi \in W_1(I)$, $I = (a,b)$ be such that

$$(\Psi,\Phi) + (\Psi',\Phi') = 0, \text{ for all } \Psi \in C_\infty(I) \cap W_1(I). \tag{1.9}$$

We have to show that

$$\Phi = 0 .$$

As a specialization of (1.9) (choosing $\varphi \in C_\infty(I)$) we obtain

$$0 = (\varphi', \Phi') + (\varphi, \Phi) = (-\varphi'' + \varphi, \Phi) ,$$

for all $\varphi \in C_\infty(I)$. We first observe that

$$-\varphi''(x) + \varphi(x) = e^{x}(e^{-2x}(e^{x}\varphi)')'.$$

This leads by similar considerations as in the proof of Lemma 1.1 to letting φ be a modified solution of

$$e^{-2x}(e^{x}\varphi)' = \int_{I\cap(-\infty,x]} e^{-t} f(t)\,dt - \int_{I\cap(-\infty,x]} e^{-t}\alpha_1(t)\,dt \int_I e^{-t} f(t)\,dt ,$$

where $\alpha_1 \in C_\infty(I)$ such that

$$\int_I e^{-t}\alpha_1(t)\,dt = 1,$$

and f an arbitrary function in $C_\infty(I)$.

One more step of the same type leads to the modified solution

$$\begin{aligned}\varphi(x) = e^{-x}\Big\{ &\int_{I\cap(-\infty,x]} e^{2t}\Big[\int_{I\cap(-\infty,t]} e^{-s} f(s)\,ds + \\ &- \int_{I\cap(-\infty,t]} e^{-s}\alpha_1(s)\,ds \int_I e^{-s} f(s)\,ds\Big]\,dt + \\ &- \int_{I\cap(-\infty,x]} e^{t}\alpha_2(t)\,dt \int_I e^{2t}\Big[\int_{I\cap(-\infty,t]} e^{-s} f(s)\,ds + \\ &- \int_{I\cap(-\infty,t]} e^{-s}\alpha_1(s)\,ds \int_I e^{-s} f(s)\,ds\Big]\,dt\Big\}\end{aligned}$$

where $\alpha_2 \in C_\infty(I)$ is such that

$$\int_I e^{t}\alpha_2(t)\,dt = 1$$

By this construction we have $\varphi \in C_\infty(I)$ and

$$-\varphi''(x) + \varphi(x) = f(x) - \alpha_1(x)\int_I e^{-s} f(s)\,ds +$$

$$+ \alpha_2(x) \int_I e^{2t} \left[\int_{I\cap(-\infty,t]} e^{-s} f(s)\, ds - \int_{I\cap(-\infty,t]} e^{-s} \alpha_1(s)\, ds \int_I e^{-s} f(s)\, ds \right] dt$$

$$- \alpha_2'(x) \int_I e^{2t} \left[\int_{I\cap(-\infty,t]} e^{-s} f(s)\, ds - \int_{I\cap(-\infty,t]} e^{-s} \alpha_1(s)\, ds \int_I e^{-s} f(s)\, ds \right] dt,$$

$$= f(x) - \alpha_1(x) \int_I e^{-s} f(s)\, ds +$$

$$- \frac{1}{2} (\alpha_2(x) - \alpha_2'(x)) \left\{ \int_I e^{t} f(t)\, dt - \int_I e^{t} \alpha_1(t)\, dt \int_I e^{-s} f(s)\, ds \right\}.$$

Thus we have for any $f \in C_\infty(I)$

$$(f, \Phi) = (\alpha_1, \Phi) \int_I e^{-s} \overline{f(s)}\, ds) +$$

$$- \frac{1}{2} ((\alpha_2 - \alpha_2'), \Phi) \int_I e^{t} \overline{f(t)}\, dt +$$

$$+ \frac{1}{2} ((\alpha_2 - \alpha_2'), \Phi) \int_I e^{t} \overline{\alpha_1(t)}\, dt \int_I e^{-s} \overline{f(s)}\, ds.$$

This implies

$$\Phi(x) = ((\alpha_1, \Phi) + \frac{1}{2} ((\alpha_2 - \alpha_2'), \Phi) \int_I e^{t} \overline{\alpha_1(t)}\, dt)\, e^{-x} - \frac{1}{2} ((\alpha_2 - \alpha_2'), \Phi)\, e^{x}.$$

In other words $\Phi = C_1\, e^{m} + C_2\, e^{-m}$. Thus from (1.9) we have

$$0 = (\varphi, \Phi) + (\varphi', \Phi') = C_1\, (\varphi + \varphi', e^{m}) + C_2\, (\varphi - \varphi', e^{-m}),$$

for all $\varphi \in C_\infty(I) \cap W_1(I)$. Letting $\varphi = e^{m}$ and $\varphi = e^{-m}$ we see that C_1, C_2 vanish identically. This in turn proves the density statement of the lemma.

□

Note that according to Lemma 1.2 we have that

$$\Phi \in C_\infty(I) \cap W_1(I) \equiv C_\infty(I) \cap H_1(I)$$

and the limit

$$\lim_{t\to t_o} \Phi(t) \text{ exists as } t \in I \text{ tends to } t_o \in \partial I \equiv \{a,b\}.$$

Thus defining $\Phi(a)$, $\Phi(b)$ in this sense Lemma 1.4 and Lemma 1.2 lead to the definition of

$$\Phi_{\restriction t} := \lim_{n\to\infty} \Phi_n(t) \quad , \quad t \in \Gamma \, , \tag{1.10}$$

for $\Phi \in W_1(I)$, with $(\Phi_n)_n \subset C_\infty(I) \cap W_1(I) \equiv C_\infty(I) \cap H_1(I)$ and $\Phi_n \to \Phi$ in $W_1(I)$ as $n \to \infty$ (compare [1], p.38).

The trace $\Phi_{\restriction t}$ is well defined, since it is independent of the choice of the approximating sequence.

We are finally in the state to give the alternative description of $D(d)$ as

<u>Lemma 1.5</u>: $$D(d) = \{\Phi \in W_1(I) \mid \Phi_{\restriction a} = \Phi_{\restriction b} = 0\} \, .$$

<u>Proof</u>: Since $D(d) \subset \{\Phi \in W_1(I) \mid \Phi_{\restriction a} = \Phi_{\restriction b} = 0\}$ is immediate from (1.8), it only remains to show that an element $\Phi \in W_1(I)$ satisfying the boundary condition

$$\Phi_{\restriction \partial I}(x) := \left\{ \begin{array}{ll} \Phi_{\restriction a} & \text{if} \quad x = a \\ \Phi_{\restriction b} & \text{if} \quad x = b \end{array} \right\} = 0 \, ,$$

is contained in $H_1(I)$. Let $X := \{\Phi \in W_1(I) \mid \Phi_{\restriction a} = \Phi_{\restriction b} = 0\}$, for sake of brevity, then we observe that according to Lemma 1.2 and the above considerations X is a closed subspace of $W_1(I)$. Application of the projection theorem to the Hilbert space X and $H_1(I)$ as a closed subspace of X yields

$$X = H_1(I) \oplus (H_1(I))^\perp \, .$$

We have to show that

$$(H_1(I))^\perp = \{0\} \, . \tag{1.11}$$

So, let $f \in (H_1(I))^\perp$, i.e.

$$0 = (\varphi' \, , \, f') + (\varphi, f) = (-\varphi'' + \varphi \, , \, f) \, ,$$

for all $\varphi \in C_\infty(I)$. As in the proof of Lemma 1.4 we see that

$$f(x) = ((\alpha_1, f) + \frac{1}{2}((\alpha_2 - \alpha_2'), f) \int_I e^t \, \overline{\alpha_1(t)} \, dt) \, e^{-x} - \frac{1}{2}((\alpha_2 - \alpha_2'), f) \, e^x$$

As an element of X the function f has to satisfy the boundary conditions as well. This is only possible if $(\alpha_1, f) = ((\alpha_2 - \alpha_2'), f) = 0$, i.e. $f \equiv 0$. □

Another description of $D(d)$ may be obtained from the following concept of "weak" boundary condition. The space

$$\overset{\circ}{W}_1(I) := \{\Phi \in W_1(I) \mid (\Psi, \Phi') + (\Psi', \Phi) = 0 \text{ for all } \Psi \in W_1(I)\}$$

is a closed subspace of $W_1(I)$ as well.
In fact, we have

<u>Lemma 1.6</u>: $\overset{\circ}{W}_1(I) = \overset{\circ}{H}_1(I)$.

<u>Proof</u>: $\overset{\circ}{H}_1(I) \subset \overset{\circ}{W}_1(I)$ is obvious. Let $\Phi \in \overset{\circ}{W}_1(I)$ then

$$(\Psi,\Phi') + (\Psi',\Phi) = 0 \text{ for all } \Psi \in C_\infty(\bar{I}) \text{ .} \tag{1.13}$$

On the other hand, an approximation argument shows that

$$\int_I \overline{\Psi(t)}\, \frac{d}{dt}\Phi(t)\, dt + \int_I \frac{d}{dt}\overline{\Psi(t)} \cdot \Phi(t)\, dt = \overline{\Psi(b)}\, \Phi_{\restriction b} - \overline{\Psi(a)}\, \Phi_{\restriction a}, \tag{1.14}$$

for $\Psi \in C_\infty(\bar{I})$. Comparison of (1.13), (1.14) shows

$$\Phi_{\restriction b} = 0 \text{ , } \Phi_{\restriction a} = 0 \text{ .}$$

□

Now we have three characterizations of $D(d)$:

$$D(d) = \overset{\circ}{H}_1(I) = \overset{\circ}{W}_1(I) = \{\Phi \in W_1(I) \mid \Phi_{\restriction b} = \Phi_{\restriction a} = 0 \} \text{ .}$$

We are now ready for considering selfadjoint extensions of d. By a simple transformation we may assume w. l. o. g. that $I = (-1/2 \text{ , } 1/2)$.

The general theory of selfadjoint extensions (see appendix,, Theorem A.7) tells us that we have to investigate the Cayley transform $C(d)$ of the closed symmetric operator d.

We recall: $C(A) = (A+i)\ (A-i)^{-1}$ is isometric from the range $R(A-i)$ of $A-i$ to $R(A+i)$

$$A < B \text{ if } C(A) < C(B) \text{ , ("}A < B\text{" read "A is continued by B")}$$

$$D(A) = R(1 - C(A)) \text{ ,} \tag{1.15}$$

$$D(B) = D(A) \oplus (1 - C(B))\, N(A^* + i) \text{ ,}$$

$$(1 - C(B))\, (N(A^*+i) \cap R(B-i)) = N(A^*-i) \cap R(B+i)$$

$$D(1-C(A)) = D(C(A)) = R(A-i) \text{ ,}$$

for A,B any closed, symmetric operators. In particular,

B selfadjoint iff C(B) unitary.

In order to find selfadjoint extensions $\tilde{d}$ of d it is convenient to search for unitary extensions $\tilde{U}$ of the Cayley transform of d and then to define in accordance with (1.15)

$$D(\tilde{d}) := D(d) \oplus (1-\tilde{U})\, N(d^* + i)) \ ,$$

$$\tilde{d}\Phi := - i\, (1+\tilde{U})\, (1-\tilde{U})^{-1}\, \Phi \text{ for all } \Phi \in D(\tilde{d}).$$

Any selfadjoint extension may be described by specifying $\tilde{U}$ on $N(d^*+i)$ (observe $\tilde{U}x := U(d)x$ for $x \in R(d-i)$). In our particular case we know that $e^{\pm 2\pi m}$ is the only element in $N(d^* \pm i)$. Thus, our possibilities are limited to conditions of the type

$$\tilde{U}(e^{+2\pi m}) = \alpha\, e^{-2\pi m} \text{ for some } \alpha \in \mathbf{C} \ ,$$

such that

$$|\alpha|\ \|e^{-2\pi m}\| = \|e^{+2\pi m}\| \ ,$$

i.e.

$$|\alpha| = 1.$$

If we denote the corresponding selfadjoint extension by d_α we see that $\Phi \in D(d_\alpha)$ is characterized by the boundary conditions

$$\Phi_{\restriction -1/2} = c(e^{-\pi} - \alpha\, e^{\pi}) \ ,$$

$$\Phi_{\restriction +1/2} = c(e^{\pi} - \alpha\, e^{-\pi}) \ , \ c \in \mathbf{C} \ ,$$

or

$$(e^{\pi} - \alpha\, e^{-\pi})\Phi_{\restriction -1/2} - (e^{-\pi} - \alpha\, e^{\pi})\Phi_{\restriction +1/2} = 0 \ . \tag{1.16}$$

The choice

$$\alpha = -1$$

leads to the periodicity boundary condition:

$$\Phi_{\restriction -1/2} = \Phi_{\restriction +1/2} \tag{1.17}$$

It is easy to see that

$$D(d_{-1}) = \{\Phi \in W_1(I) \mid \Phi_{\restriction -1/2} = \Phi_{\restriction +1/2} \}.$$

Let us restrict our attention to this particular case (We define $D^{\#} := d_{-1}$ to emphasize the importance of this particular operator). To describe the spectral properties of the selfadjoint extension $D^{\#}$ of d we have

Theorem 1.1: $D^{\#}$ *has a compact resolvent.*

Proof: Since $D(D^{\#}) \subset H_1(I)$ we have a $C_{1/2}$-trace of any $\Phi \in D(D^{\#})$ in Γ. Let

$$(\Phi_n)_n \subset H_1(I)$$

be a bounded sequence. Then according to Lemma 1.2, Lemma 1.3

$$(\Phi_n)_n \text{ bounded in } C_{1/2}(I) .$$

In particular, $(\Phi_n)_n$ is equi-continuous. By the Arzela-Ascoli theorem we have a subsequence converging in $C_0(I)$ and thus in $L_2(I)$. This way we have obtained the compact imbedding result

$$H_1(I) \text{ compactly imbedded in } L_2(I) . \tag{1.18}$$

A-forteriori we conclude that $D(D^{\#})$ is compactly imbedded in $L_2(I)$. This implies the compactness of $(D^{\#}-i)^{-1}$.

□

Corollary 1.1: *The spectrum* $\sigma(D^{\#})$ *of* $D^{\#}$ *is discrete.*

(See appendix, Theorem A.10).

In order to find the eigenvalues and eigensolutions, we observe that ($\lambda \in \mathbb{R}$)

$$\Phi \in N(D^{\#}-\lambda) \text{ iff } ((D-\lambda)\Psi,\Phi) = 0 \text{ for all } \Psi \in C_\infty(I) \text{ and } \Phi_{\restriction 0} = \Phi_{\restriction 2\pi} .$$

Using similar considerations as before, it is clear that the classical solutions of

$$D\,\Phi - \lambda\,\Phi = 0 ,$$

are the only possible eigensolutions (for the technique see proof of Lemma 1.4), i.e.

$$\Phi = C\, e^{2\pi i \lambda m} \text{ for some } C \in \mathbb{C} .$$

From the periodicity properties of e^z we conclude

$$\sigma(D^{\#}) = \mathbb{Z} ,$$

(here and in the following $\mathbb{Z}$ denotes the set of integers) and the normalized eigenfunctions are

$$\Phi_k = e^{2\pi i k m} , \ k \in \mathbb{Z} .$$

The spectral resolution $(\Pi(\lambda))_{\lambda\in\mathbb{R}}$ of $D^{\#}$ therefore takes the form (in the sense of strong convergence)

$$\Pi(\lambda)f = \sum_{\mu\in(-\infty,\lambda]\cap\sigma(D^{\#})} P_\mu f \ , \tag{1.19}$$

P_μ denotes the orthogonal projection on the eigenspace $N(D^{\#}-\mu)$, where $\mu \in \sigma(D^{\#})$:

$$P_\mu f = \int_{\mathbb{R}} e^{-2\pi i\mu t} f(t)\,dt\; e^{2\pi i\mu m}, \quad \mu \in \mathbb{Z} .$$

This may be written as

$$P_\mu f = (e^{2\pi i\mu m}, f)\; e^{2\pi i\mu m}, \quad \mu \in \mathbb{Z} . \tag{1.20}$$

Remark 1.3: *The somewhat strange looking notational idea of introducing a multiplication operator to describe functions has been used to avoid the possible confusion between free and bound variables. It is sometimes customary to use a dot for the same purpose.*
A careful distinction shows its virtue in the present situation. To give an example; $(e^{2\pi i\lambda x}, f)$ *could mean a function of* λ *or of* x *alone or a function of both variables. The previously introduced notation clarifies the situation suitably.*

For $\lambda \to +\infty$ in the equation

$$\|\Pi(\lambda)f\|^2 = \sum_{\mu\in(-\infty,\lambda]\cap\mathbb{Z}} \|P_\mu f\|^2 = \sum_{\substack{\mu<\lambda \\ \mu\in\mathbb{Z}}} |(e^{2\pi i\mu m}, f)|^2$$

we obtain Parseval's equality

$$\|f\|^2 = \sum_{\mu\in\mathbb{Z}} |(e^{2\pi i\mu m}, f)|^2 , \quad f \in L_2(I) . \tag{1.21}$$

Equation (1.20) establishes the unitary equivalence between $L_2(I)$ and the sequence space ℓ_2 (i.e. the set of all sequences $x = (x_i)_{i\in\mathbb{Z}}$ of complex numbers such that $\sum_{i\in\mathbb{Z}} |x_i|^2 < \infty$, equipped with the inner product $\sum_{i\in\mathbb{Z}} \overline{x_i}\, y_i$, $x,y \in \ell_2$). This unitary relationship is the Fischer-Riesz theorem. With the Dirac measure

$$\delta_{\mathbb{Z}}(J) = \sum_{\mu\in J\cap\mathbb{Z}} 1 \ , \ J \text{ an interval in } \mathbb{R} \ ,$$

ℓ_2 may be written as $L_2(\mathbb{R},\delta_{\mathbb{Z}})$. So, we may define the unitary equivalence

$$V: L_2(I) \longrightarrow L_2(\mathbb{R},\delta_{\mathbb{Z}}) \ , \qquad f \longrightarrow \hat{f} \tag{1.22}$$

where

$$\hat{f}(t) := \begin{cases} (e^{2\pi i t m}, f) \ , & t \in \mathbb{Z} \\ 0 & \text{otherwise} \end{cases} .$$

Since V maps $D^{\#}$ into the selfadjoint multiplication operator $\boldsymbol{m}$,

$$\boldsymbol{m} : V\, H_1(I) \subset L_2(\mathbb{R},\delta_{\mathbb{Z}}) \longrightarrow L_2(\mathbb{R},\delta_{\mathbb{Z}}) \ , \qquad \varphi \longrightarrow m\, \varphi ,$$

we have

$$V(D^{\#}\Phi)(\lambda) = \lambda\, V(\Phi)(\lambda) = \boldsymbol{m}\, V(\Phi)\ (\lambda) \ , \ \lambda \in \mathbb{R} \ ,$$

or

$$V\, D^{\#} = \boldsymbol{m}\, V.$$

Thus, V is a spectral representation of $D^{\#}$. The spectrum of $D^{\#}$ has multiplicity one and is therefore simple. Note that $\boldsymbol{m}$ is a selfadjoint operator on ℓ_2. From (1.9) we obtain for $\lambda \to \infty$ the <u>Fourier series</u> for f:

$$f = \sum_{\mu\in\mathbb{Z}} (e^{2\pi i \mu \boldsymbol{m}}, f) e^{2\pi i \mu \boldsymbol{m}} \ , \ f \in L_2(I) \ ,$$

We collect the results obtained so far in

<u>Theorem 1.2</u>: $D^{\#}$ *is a selfadjoint operator with spectrum*

$$\sigma(D^{\#}) = \mathbb{Z} \ .$$

The eigenspaces are 1-dimensional

$$N(D^{\#}-k) = \operatorname{span}(e^{2\pi i k \boldsymbol{m}}) , \ k \in \mathbb{Z} \ ,$$

and the spectral family has the form

$$\Pi(\lambda) f = \sum_{\substack{\mu\le\lambda \\ \mu\in\mathbb{Z}}} (e^{2\pi i \mu \boldsymbol{m}}, f)\ e^{2\pi i \mu \boldsymbol{m}} .$$

The unitary mapping V *defined in* (1.21) *is a spectral representation for the operator* $D^{\#}$.

1.1.2 Selfadjoint realization of the operator D in the case of an unbounded interval. The Fourier transform.

The closure $\bar{d}$ of d is selfadjoint in this case and we shall denote it by D. Similar as in 1.1.1 we may characterize the domain of D by

$$D(D) = H_1(\mathbb{R}) := \overline{C_\infty(\mathbb{R})}^{\|\cdot\|_1}$$

and

$$D(D) = W_1(\mathbb{R}) = \{\Phi \in L_2(\mathbb{R}) \mid \Phi' \in L_2(\mathbb{R})\} .$$

Since $e^{2\pi i\lambda m}$, $\lambda \in \mathbb{R}$, are the only candidates for eigensolutions (see paragraph 1.1.1), it is obvious that the point spectrum $P\sigma(A) = \emptyset$. That the spectrum of D is all of $\mathbb{R}$ may be concluded from Stone's formula (see below) or shown directly by constructing a singular sequence.

Let

$$u_\varepsilon(x) := \Phi(x)\, e^{2\pi i\lambda x}\, e^{-2\pi\varepsilon x} , \quad \Phi' \in C_\infty(\mathbb{R}) , \quad \Phi(+\infty) = 1 , \quad \Phi(-\infty) = 0 ,$$

$\lambda \in \mathbb{R}$, $\varepsilon > 0$, $\Phi \geq 1$ on $\mathbb{R}^+$, then

$$Du_\varepsilon = \frac{1}{2\pi i}\Phi'\, e^{2\pi i\lambda m}\, e^{-2\pi\varepsilon m} + \lambda\, u_\varepsilon + i\varepsilon\, u_\varepsilon ,$$

or

$$Du_\varepsilon - \lambda u_\varepsilon = \frac{1}{2\pi i}\Phi'\, e^{2\pi i\lambda m}\, e^{-2\pi\varepsilon m} + i\varepsilon\, u_\varepsilon =: f_\varepsilon .$$

$\|f_\varepsilon\|$ is uniformly bounded as $\varepsilon \to 0$, but

$$\|u_\varepsilon\|^2 \geq \int_{\mathbb{R}^+} e^{-4\pi\varepsilon x}\, dx = \frac{1}{4\varepsilon\pi} \longrightarrow \infty \text{ as } \varepsilon \longrightarrow 0 .$$

Thus

$$\sigma(D) = \mathbb{R} ,$$

and since there is no point spectrum, the spectrum is continuous:

$$C\sigma(D) = \sigma(D) = \mathbb{R} .$$

A Green's function may be defined by

$$g_\lambda(x) := 2\pi i \begin{cases} e^{2\pi i\lambda x} & \text{if } x \geq 0 \\ 0 & \text{if } x < 0 \end{cases}$$

if Im $\lambda > 0$, and

$$g_\lambda(x) := 2\pi i \begin{cases} 0 & \text{if } x \geq 0 \\ -e^{2\pi i \lambda x} & \text{if } x < 0 \end{cases}$$

if $\operatorname{Im} \lambda < 0$.

For $\lambda \in \mathbb{C}\backslash\mathbb{R}$

$$g_\lambda \in L_2(\mathbb{R}) \ .$$

For $f \in L_2(\mathbb{R})$ with compact support an elementary calculation shows that the resolvent of D can be represented as

$$((D - \lambda)^{-1} f)(x) = \int_{\mathbb{R}} g_\lambda(x-y) \ f(y) \ dy \ . \tag{1.23}$$

Representation (1.23) will be used to calculate the spectral family via Stone's formula (appendix, Theorem A.8):

$$\left(g \ , \ \left(\frac{\Pi(\lambda_2^+) + \Pi(\lambda_2^-)}{2} - \frac{\Pi(\lambda_1^+) + \Pi(\lambda_1^-)}{2} \right) f \right) =$$

$$= (g \ , \ (\Pi(\lambda_2) - \Pi(\lambda_1) \ ) \ f \ )$$

$$= \lim_{\varepsilon \to 0+} \frac{1}{2\pi i} \int_{\lambda_1}^{\lambda_2} (g \ , \ ((D - (\mu + i\varepsilon))^{-1} - (D - (\mu - i\varepsilon))^{-1} \ ) \ f) \ d\mu$$

$$= \lim_{\varepsilon \to 0+} \frac{1}{2\pi i} \int_{\lambda_1}^{\lambda_2} \int_{\mathbb{R}} \overline{g(t)} \int_{\mathbb{R}} (g_{\lambda+i\varepsilon}(t-s) - g_{\lambda-i\varepsilon}(t-s)) \ f(s) \ ds \ dt \ d\mu$$

$$= \frac{1}{2\pi i} \int_{\lambda_1}^{\lambda_2} \int_{\mathbb{R}} \int_{\mathbb{R}} \overline{g(t)} \ (g_{\lambda+i0}(t-s) - g_{\lambda-i0}(t-s)) \ f(s) \ ds \ dt \ d\mu$$

$$= \int_{\lambda_1}^{\lambda_2} \int_{\mathbb{R}} \int_{\mathbb{R}} \overline{g(t)} \ e^{2\pi i \lambda t} \cdot e^{-i\lambda s} \ f(s) \ ds \ dt \ d\mu$$

$$= \int_{\lambda_1}^{\lambda_2} \int_{\mathbb{R}} e^{-2\pi i \lambda t} \ g(t) \ dt \cdot \int_{\mathbb{R}} e^{-2\pi i \lambda s} \ f(s) \ ds \ d\mu$$

for $f,g \in L_2(\mathbb{R})$ with compact support (i.e. supp (f) $\subset\subset \mathbb{R}$, supp (g) $\subset\subset \mathbb{R}$).
Setting

$$F_\sigma(f)(\lambda) := \sigma \int_{\mathbb{R}} e^{-2\pi i\lambda t} f(t)\, dt, \quad f \in L_2(\mathbb{R}) \;,\; \text{supp}\,(f) \subset\subset \mathbb{R} \;, \qquad (1.24)$$

with, $\sigma \in \mathbb{C}$, $|\sigma| = 1$, we obtain

$$(f,(\Pi(\lambda_2) - \Pi(\lambda_1))\, f) = \|(\Pi(\lambda_2) - \Pi(\lambda_1))\, f\|^2 = \int_{\lambda_1}^{\lambda_2} |F_\sigma(f)(\lambda)|^2\, d\lambda \;.$$

For $\lambda_2 \to +\infty$, $\lambda_1 \to -\infty$ we get from the spectral theorem Parseval's equality

$$\|f\|^2 = \int_{\mathbb{R}} |F_\sigma(f)(\lambda)|^2\, d\lambda = \|F_\sigma(f)\|^2 \;,\; f \in L_2(I) \;,\; \text{supp}\,(f) \subset\subset \mathbb{R} \;.$$

This shows that F_σ can be extended by continuity to an isometry defined for $f \in L_2(I)$. The reason for leaving a parameter σ in the definition above will become clear later (see 1.1.2.3). It is usually assumed that $\sigma = 1$ and the extension is called the <u>Fourier transform</u> (or Fourier-Plancherel transform):

$$\mathcal{F}(f)(\lambda) \equiv F_1(f)(\lambda) = \int_{\mathbb{R}} e^{-2\pi i\lambda x} f(x)\, dx \;,$$

where the integral should be understood in the sense of this extension. To emphasize the analogy to the Fourier coefficients of the previous chapter we might use the suggestive notation

$$(e^{2\pi i\lambda m}, f) \equiv \mathcal{F}(f)(\lambda) .$$

We shall extend the meaning of the $L_2(\mathbb{R})$ inner product properly in section 1.1.2.2.

Since $\mathcal{F}^*(f)(\lambda) = \mathcal{F}(f)(-\lambda)$ (see (1.28)) it is obvious that $\mathcal{F}$ is onto (note $N(\mathcal{F}^*) = \{0\}$). Furthermore,

$$\mathcal{F}(Du)(\lambda) = \lambda\, \mathcal{F}(u)(\lambda) \text{ for } u \in D(A),$$

or

$$\mathcal{F}\, D = \boldsymbol{m}\, \mathcal{F}. \qquad (1.25)$$

Relation (1.25) shows that $\mathcal{F}$ is a spectral representation of D. We remark that in this case $\boldsymbol{m}$ is a selfadjoint operator on $L_2(\mathbb{R})$. We state our results as

<u>Theorem 1.3</u>: D *is a selfadjoint operator with spectrum* $\sigma(D) = C\sigma(D) = \mathbb{R}$. *The spectrum is simple. The* <u>*Fouriertransform*</u> $\mathcal{F}$ *is a spectral representation of* D.

1.1.2.1 Basic properties of the Fourier transform.

To see that the Fourier transform $\mathcal{F}$ is onto we have already used an interesting feature of $\mathcal{F}$. Let $\mathcal{C}_{\alpha,\beta}$, $\alpha,\beta \in \mathbb{R}$, $\alpha \neq 0$, be a linear coordinate transformation such that

$$(\mathcal{C}_{\alpha,\beta} f)(x) = f(\alpha x+\beta).$$

Note that

$$\|\mathcal{C}_{\alpha,\beta} f\| = \frac{1}{\sqrt{|\alpha|}} \|f\|, \text{ for all } f \in L_2(\mathbb{R}),$$

and in particular

$$\mathcal{C}^*_{\alpha,\beta} = \frac{1}{\sqrt{|\alpha|}} \mathcal{C}_{1/\alpha,-\beta/\alpha} = \frac{1}{\sqrt{|\alpha|}} (\mathcal{C}_{\alpha,\beta})^{-1}. \tag{1.26}$$

As a special case we see that apparently $\mathcal{C}_{\pm 1,\beta}$ is unitary. An elementary calculation yields

$$\begin{aligned} \mathcal{C}_{\alpha,\beta} \mathcal{F} &= \frac{1}{\sqrt{|\alpha|}} \mathcal{F} e^{-2\pi i(\beta/\alpha)\mathbf{m}} \mathcal{C}_{1/\alpha,0}, \\ &= \frac{1}{\sqrt{|\alpha|}} \mathcal{F}^* e^{2\pi i(\beta/\alpha)\mathbf{m}} \mathcal{C}_{-1/\alpha,0} \quad \text{on } L_2(\mathbb{R}). \end{aligned} \tag{1.27}$$

Here $\mathbf{m}$ is again the multiplication operator corresponding to multiplication by the argument, so

$$(e^{\pm 2\pi i(\beta/\alpha)\mathbf{m}} f)(x) = e^{2\pi i(\beta/\alpha)x} f(x). \tag{1.28}$$

As an immediate consequence of (1.27) we obtain

$$\begin{aligned} \mathcal{F}^* \mathcal{C}_{\alpha,\beta} &= \frac{1}{\sqrt{|\alpha|}} e^{-2\pi i(\beta/\alpha)\mathbf{m}} \mathcal{C}_{1/\alpha,0} \mathcal{F}^*, \\ \mathcal{F} \mathcal{C}_{\alpha,\beta} &= \frac{1}{\sqrt{|\alpha|}} e^{+2\pi i(\beta/\alpha)\mathbf{m}} \mathcal{C}_{-1/\alpha,0} \mathcal{F}^* \quad \text{on } L_2(\mathbb{R}). \end{aligned} \tag{1.29}$$

The choice $\alpha = -1$, $\beta = 0$ yields

$$\mathcal{C}_{-1,0} \mathcal{F} = \mathcal{F} \mathcal{C}_{-1,0} = \mathcal{F}^* \text{ on } L_2(\mathbb{R}), \tag{1.30}$$

which was used in the above to see that $N(\mathcal{F}^*) = \{0\}$. Note that $\mathcal{C}_{-1,0}$ is a selfadjoint isometry on $L_2(\mathbb{R})$. The operators

$$\text{sym} = \frac{1}{2}(1 + \mathcal{C}_{-1,0}), \quad \text{asym} = \frac{1}{2}(1 - \mathcal{C}_{-1,0}), \tag{1.31}$$

produce the even and odd part of a function, respectively. They are orthogonal projections of $L_2(\mathbb{R})$:

$$1 - \text{sym} = \text{asym}.$$

The Fourier transform interacts with these projections according to (1.30) in the following way:

$$\mathcal{F}\ \mathrm{sym} = \frac{1}{2}\ (\mathcal{F} + \mathcal{F}^{*})\ =:\ \mathrm{Re}(\mathcal{F}) = \mathrm{sym}\ \mathcal{F}\ ,$$
$$\mathcal{F}\ \mathrm{asym} = \frac{1}{2}\ (\mathcal{F} - \mathcal{F}^{*})\ =:\ i\ \mathrm{Im}(\mathcal{F}) = \mathrm{asym}\ \mathcal{F}\ . \tag{1.32}$$

$\mathrm{Re}(\mathcal{F})$, $\mathrm{Im}(\mathcal{F})$ are known as (double-sided) Fourier sine and Fourier cosine transform. They are real transforms, i.e. they transform real functions into real functions. We shall reconsider these transforms in a later section of this monograph.

With respect to (1.30) we also note that

$$\mathcal{F}\ \mathcal{F} = \mathcal{F}^{*}\ \mathcal{F}^{*} = \ell_{-1,0}. \tag{1.33}$$

This implies that sym $(1 \pm \mathcal{F})$ is projection on the eigenspace $N(\mathcal{F} \pm 1)$ of $\mathcal{F}$ for the eigenvalue ∓ 1 (compare [34], p. chapter 9, where also an alternative description of $N(\mathcal{F} - 1)$ is given).

The transformation $\ell_{1,-h}$ is called a shift operation and denoted by σ_h , $h \in \mathbb{R}$. As all $\ell_{\pm 1,\beta}$, $\beta \in \mathbb{R}$, σ_h is a unitary mapping from $L_2(\mathbb{R})$ onto $L_2(\mathbb{R})$. We have

$$\sigma_h\ \mathcal{F} = \mathcal{F}\ e^{2\pi i h \boldsymbol{m}} \text{ on } L_2(\mathbb{R}) \tag{1.34}$$

and

$$\mathcal{F}\ \sigma_h = e^{-2\pi i h \boldsymbol{m}}\ \mathcal{F} \text{ on } L_2(\mathbb{R}). \tag{1.35}$$

Using (1.29) we can get analogous properties for $\mathcal{F}^{*}$. By (1.29) it is clear that we have from (1.23)

$$\mathcal{F}^{*}\ D = -\boldsymbol{m}\ \mathcal{F}^{*} \text{ on } H_1(\mathbb{R}), \tag{1.36}$$

or

$$D\ \mathcal{F} = -\mathcal{F}\ \boldsymbol{m} \text{ on } \mathcal{F}\ H_1(\mathbb{R}). \tag{1.37}$$

Thus, if $\Psi \in H_1(\mathbb{R}) \cap \mathcal{F}\ H_1(\mathbb{R})$ we have

$$\mathcal{F}\ (D + i\ \boldsymbol{m})\ \Psi = (\boldsymbol{m} - i\ D)\ \mathcal{F}\ \Psi = -\ i\ (D + i\ \boldsymbol{m})\ \mathcal{F}\ \Psi. \tag{1.38}$$

Now let

$$\gamma = e^{-\pi\ \boldsymbol{m}^2},$$

then we have for $\Phi \in C_\infty(\mathbb{R})$

$$(D + i\ \boldsymbol{m})\ \Phi = (\gamma^{-1}\ D\ \gamma)\ \Phi. \tag{1.39}$$

We shall denote the differential operator $(\gamma^{-1} D \gamma)$ by $\mathcal{D}$

$$\mathcal{D} \Phi = (D + i \, m) \Phi = (\gamma^{-1} D \gamma) \Phi \text{ for } \Phi \in C_\infty(\mathbb{R}). \tag{1.40}$$

With this we have the following interesting property of the Fourier transform

$$i \, \mathcal{F} \mathcal{D} = \mathcal{D} \mathcal{F} \text{ on } H_1(\mathbb{R}) \cap \mathcal{F} H_1(\mathbb{R}). \tag{1.41}$$

The formal adjoint $\mathcal{D}^*$ of $\mathcal{D}$ is apparently

$$\mathcal{D}^* \Phi = (D - i \, m) \Phi .$$

$\mathcal{D}^*$ is known as tempered derivative (see [4]). In the next paragraph we shall show that $H_1(\mathbb{R}) \cap \mathcal{F} H_1(\mathbb{R})$ are suitable domains for both $\mathcal{D}$ and $\mathcal{D}^*$.

<u>Remark 1.4</u>: *The Gauss function* $\gamma(x) = e^{-\pi x^2}$ *used to define the operator* γ *has the remarkable property that it is an eigenfunction of* $\mathcal{F}$ *(note:* $\gamma \in L_2(\mathbb{R})$*) for eigenvalue 1:*

$$\mathcal{F} \gamma = \gamma.$$

From (1.40) *we find that*

$$\mathcal{F} \mathcal{D} \gamma = - i \, \mathcal{D} \gamma.$$

Note that

$$\mathcal{D}^* \gamma = 0, \; \mathcal{D} \gamma = 2 \, D \gamma = 2 \, i \, m \gamma ,$$

this implies

$$\mathcal{F} D \gamma = - i \, D \gamma$$

and

$$\mathcal{F} D^2 \gamma = - D^2 \gamma, \; \mathcal{F} D^3 \gamma = i \, D^3 \gamma, \text{ etc. },$$

as well as

$$\mathcal{F} \mathcal{D}^2 \gamma = - \mathcal{D}^2 \gamma, \; \mathcal{F} \mathcal{D}^3 \gamma = i \, \mathcal{D}^3 \gamma, \text{ etc. }.$$

<u>1.1.2.2 Tempered Sobolev spaces</u>.

To develop a proper setting for the operator $\mathcal{D}$ we start with the observation that for $\Phi \in C_\infty(\mathbb{R})$ we have

$$\|\mathcal{D}\Phi\|^2 = \|(D+i m)\Phi\|^2 = \|D\Phi\|^2 + \|m\Phi\|^2 + 2 \, \mathrm{Re}\, (i \, (D\Phi, m\Phi)),$$

$$= \|D\Phi\|^2 + \|\mathfrak{m}\Phi\|^2 - \frac{1}{\pi} \operatorname{Re} (\Phi', \mathfrak{m}\Phi),$$

$$= \|D\Phi\|^2 + \|\mathfrak{m}\Phi\|^2 - \frac{1}{\pi} \int_{\mathbb{R}} \frac{t}{2} \frac{d}{dt} |\Phi(t)|^2 \, dt, \tag{1.42}$$

$$= \|D\Phi\|^2 + \|\mathfrak{m}\Phi\|^2 + \frac{1}{2\pi} \|\Phi\|^2.$$

Apparently,

$$\|\Phi\|_{\mathcal{S},1} = \|\mathcal{D}\Phi\| \text{ for } \Phi \in C_\infty(\mathbb{R}) \tag{1.43}$$

defines a norm. Now, let the completion of $C_\infty(\mathbb{R})$ with respect to this norm be denoted by $\mathcal{S}_1$. By definition $\mathcal{S}_1$ is a Hilbert space with inner product

$$(\Phi,\Psi)_{\mathcal{S},1} = (\mathcal{D}\Phi, \mathcal{D}\Psi) \text{ for } \Phi,\Psi \in \mathcal{S}_1. \tag{1.44}$$

Remark 1.5: *The above calculation showing that* $\|\bullet\|_{\mathcal{S},1}$ *is a norm also yields*

$$\operatorname{Re}(i(D\,\Phi, \mathfrak{m}\,\Phi)) = \frac{1}{4\pi} \|\Phi\|^2.$$

This can be estimated by using the Cauchy-Schwarz inequality as

$$\frac{1}{4\pi} \|\Phi\|^2 \le |\operatorname{Re}(i(D\,\Phi, \mathfrak{m}\,\Phi))| \le |(D\,\Phi, \mathfrak{m}\,\Phi)| \le \|\mathfrak{m}\Phi\| \, \|D\Phi\| \text{ for } \Phi \in \mathcal{S}_1 .$$

This estimate is known as 'Heisenberg's uncertainty principle'. Equality holds iff $\operatorname{Re}(D\Phi, \mathfrak{m}\Phi) = 0$ *and* $D\Phi = \beta\, \mathfrak{m}\Phi$ *for some* $\beta \in \mathbb{R}$*. Together with the fact that* $\Phi \in \mathcal{S}_1$ *we see that* $\beta \in \mathbb{R} + i\,\mathbb{R}^+$*. Thus, the only real elements in* $\mathcal{S}_1$ *that yield equality in the Heisenberg uncertainty estimate are of the form*

$$\mathfrak{e}_{\alpha,0}\, \gamma \text{ for } \alpha \in \mathbb{R},\ \alpha \neq 0,$$

(compare the survey article [29], p.25 pp).

The following lemma characterizes the elements in $\mathcal{S}_1$.

Lemma 1.6: *We have*

$$\mathcal{S}_1 = H_1(\mathbb{R}) \cap \mathcal{F} H_1(\mathbb{R}) = \{ f \in H_1(\mathbb{R}) \mid \mathfrak{m}\, f \in L_2(\mathbb{R}) \}.$$

In particular functions in $C_\infty(\mathbb{R})$ *are dense in* $H_1(\mathbb{R}) \cap \mathcal{F} H_1(\mathbb{R})$.

Proof: For the proof we shall employ the orthogonal projection technique used in the proof of Lemma 1.5. Norm and inner product defined in (1.43), (1.44) extend naturally to elements in $H_1(\mathbb{R}) \cap \mathcal{F} H_1(\mathbb{R})$ which thus becomes a Hilbert space. Since $\mathcal{S}_1$ is certainly contained in $H_1(\mathbb{R})$ and $\mathcal{F} H_1(\mathbb{R})$ and is a closed subspace with respect to this norm we have

$$H_1(\mathbb{R}) \cap \mathcal{F} H_1(\mathbb{R}) = \mathcal{S}_1 \oplus \mathcal{S}_1^{\perp}.$$

We shall prove

$$\mathcal{S}_1^{\perp} = \{0\}. \qquad (1.45)$$

First we see that f $\in \mathcal{S}_1^{\perp}$ iff

$$(\mathcal{D}^* \mathcal{D} \varphi, f) = 0 \text{ for all } \varphi \in C_\infty(\mathbb{R}). \qquad (1.46)$$

We shall procede as in the proofs of Lemma 1.1 and Lemma 1.5 .

Let

$$\varphi(x) = -4\pi \gamma^{-1}(x) \{ \int_{(-\infty,x]} \gamma^2(t) [\int_{(-\infty,t]} \gamma^{-1}(s) \Phi(s) ds +$$

$$- \int_{(-\infty,t]} \gamma^{-1}(s) \alpha_1(s) ds \int_{\mathbb{R}} \gamma^{-1}(s) \Phi(s) ds] dt +$$

$$- \int_{(-\infty,x]} \gamma(t) \alpha_2(t) dt \int_{\mathbb{R}} \gamma^2(t) [\int_{(-\infty,t]} \gamma^{-1}(s) \Phi(s) ds$$

$$- \int_{(-\infty,t]} \gamma^{-1}(s) \alpha_1(s) ds \int_{\mathbb{R}} \gamma^{-1}(s) \Phi(s) ds] dt \}$$

$$= - 4\pi \gamma^{-1}(x) \{ \int_{(-\infty,x]} \gamma^2(t) [\int_{(-\infty,t]} \gamma^{-1}(s) \Phi(s) ds +$$

$$- \int_{(-\infty,t]} \gamma^{-1}(s) \alpha_1(s) ds \int_{\mathbb{R}} \gamma^{-1}(s) \Phi(s) ds] dt +$$

$$+ \int_{(-\infty,x]} \gamma(t) \alpha_2(t) dt [\int_{\mathbb{R}} \int_{(-\infty,t]} \gamma^2 (s) ds \gamma^{-1}(t) \Phi(t)$$

$$- \int_{(-\infty,t]} \gamma^2 (s) ds \gamma^{-1}(t) \alpha_1(t) \int_{\mathbb{R}} \gamma^{-1}(s) \Phi(s) ds] dt \},$$

where α_1, $\alpha_2 \in C_\infty(\mathbb{R})$ is such that

$$\int_{\mathbb{R}} \gamma^{-1}(t) \alpha_1(t) dt = 1, \quad \int_{\mathbb{R}} \gamma(t) \alpha_2(t) dt = 1.$$

By this construction we have $\varphi \in C_\infty(\mathbb{R})$ and

$$\mathscr{D}^* \mathscr{D} \Psi = \Phi - \alpha_1 \int_{\mathbb{R}} \gamma^{-1}(s)\ \Phi(s)\ ds +$$

$$- (2\pi\ m + \alpha_2') \int_{\mathbb{R}} [\int_{(-\infty,t]} \gamma^2\ (s)\ ds\ \ \gamma^{-1}(t)\ \Phi(t) +$$

$$-\int_{(-\infty,t]} \gamma^2(s)\ ds\ \gamma^{-1}(t)\ \alpha_1(t) \int_{\mathbb{R}} \gamma^{-1}(s)\ \Phi(s)\ ds]\ dt. \tag{1.47}$$

Thus (1.46) implies

$$(\Phi, f) = (\alpha_1, f) \int_{\mathbb{R}} \gamma^{-1}(s)\ \overline{\Phi(s)}\ ds +$$

$$+ ((2\pi\ m + \alpha_2'), f) \int_{\mathbb{R}} [\int_{(-\infty,t]} \gamma^2\ (s)\ ds\ \ \gamma^{-1}(t)\ \overline{\Phi(t)} +$$

$$+(\int_{(-\infty,\bullet]} \gamma^2(s)\ ds\ \gamma^{-1}\ \alpha_1, f) \int_{\mathbb{R}} \gamma^{-1}(s)\ \overline{\Phi(s)}\ ds]\ dt$$

for all $\Phi \in C_\infty(\mathbb{R})$. This shows that

$$f = ((\alpha_1, f) + (\int_{(-\infty,\bullet]} \gamma^2(s)\ ds\ \gamma^{-1}\ \alpha_1, f))\ \gamma^{-1}(s) +$$

$$+((2\pi\ m + \alpha_2'), f) \int_{(-\infty,\bullet]} \gamma^2\ (s)\ ds\ \ \gamma^{-1}.$$

Since f is supposed to be in $L_2(\mathbb{R})$ we see that $f \equiv 0$ is the only possible solution.

□

The Hilbert space $\mathscr{S}_1$ is a suitable domain for the closed operator $\mathscr{D}$. In fact we have

$$\mathscr{D} = D + i\ \boldsymbol{m} \quad \text{on } \mathscr{S}_1.$$

To find the adjoint operator $\mathscr{D}^*$ we first observe that certainly

$$C_\infty(\mathbb{R}) \subset D(\mathscr{D}^*).$$

Since, similar to (1.42), we have

$$\|\mathscr{D}^* \Phi\|^2 = \|(D - i\boldsymbol{m})\Phi\|^2 = \|D\Phi\|^2 + \|\boldsymbol{m}\Phi\|^2 - 2\ \mathrm{Re}\ (i\ (D\Phi, \boldsymbol{m}\Phi)),$$

$$= \|D\Phi\|^2 + \|\mathfrak{M}\Phi\|^2 + \frac{1}{\pi} \mathrm{Re}\ (\Phi', \mathfrak{M}\Phi),$$

$$= \|D\Phi\|^2 + \|\mathfrak{M}\Phi\|^2 + \frac{1}{\pi} \int_{\mathbb{R}} \frac{t}{2} \frac{d}{dt} |\Phi(t)|^2 \, dt, \tag{1.48}$$

$$= \|D\Phi\|^2 + \|\mathfrak{M}\Phi\|^2 - \frac{1}{2\pi} \|\Phi\|^2,$$

it is also clear that $\mathcal{S}_1 \subset D(\mathcal{D}^*)$. Moreover, if we furnish $D(\mathcal{D}^*)$ with the inner product

$$(\mathcal{D}^*\Phi, \mathcal{D}^*\Psi) + \frac{1}{\pi} (\Phi, \Psi) \quad \text{for } \Phi, \Psi \in D(\mathcal{D}^*), \tag{1.49}$$

the domain $D(\mathcal{D}^*)$ of $\mathcal{D}^*$ becomes a Hilbert space. Then by the same reasoning as in the proof of Lemma 1.7 we are led to construct modified solutions to the differential equation

$$\mathcal{D}\, \mathcal{D}^*\, \Psi + \frac{1}{\pi} \Psi = \Phi \quad \text{for } \Phi \in C_\infty(\mathbb{R}).$$

Since

$$\mathcal{D}\, \mathcal{D}^*\, \Psi + \frac{1}{\pi} \Psi = \mathcal{D}^*\, \mathcal{D}\, \Psi \text{ on } C_\infty(\mathbb{R}), \tag{1.50}$$

we can use the same function Ψ to conclude that the ortho-complement of $C_\infty(\mathbb{R})$ in $D(\mathcal{D}^*)$ is $\{0\}$. Thus, we obtain

$$D(\mathcal{D}) = D(\mathcal{D}^*) = \mathcal{S}_1, \tag{1.51}$$

and

$$(\,\cdot\,,\,\cdot\,)_{\mathcal{S},1} = (\mathcal{D}^*\,\cdot\,,\ \mathcal{D}^*\,\cdot\,) + \frac{1}{\pi}(\,\cdot\,,\,\cdot\,). \tag{1.52}$$

The formal differential operator corresponding to $\mathcal{D}^*$ is called tempered derivative (see [4] p.161 pp). We now procede to consider higher order derivatives.

$$\mathcal{S}_k \text{ denotes the completion of } C_\infty(\mathbb{R}) \tag{1.53}$$

with respect to the norm induced by the inner product

$$(\cdot\,,\,\cdot)_{\mathcal{S},k} = (\mathcal{D}^k\,\cdot\,,\ \mathcal{D}^k\,\cdot\,),\ k \in \mathbb{N}. \tag{1.54}$$

For convenience we let

$$\mathcal{S}_0 := L_2(\mathbb{R}). \tag{1.55}$$

Lemma 1.7: *The Hilbert space $\mathcal{S}_k$ is continuously imbedded into $\mathcal{S}_{k-1}$,* $k \in \mathbb{N}$. *The Hilbert space $\mathcal{S}_k$ can be described as*

$$\mathcal{S}_k = \{\varphi \in L_2(\mathbb{R}) \mid \mathcal{D}^k \varphi \in L_2(\mathbb{R})\}, \quad k \in \mathbb{N}..$$

Proof: The first statement is immediate from (1.42) (applied to $\Phi = \mathcal{D}^{k-1}\varphi$). The second assertion is a density result that can be shown in the same way as Lemma 1.6. To demonstrate this conveniently we introduce a formalism that handles the construction of a modified solution in a concise way.

Let

$$J\,\Phi\,(x) := 2\pi i \left\{ \int_{(-\infty,x]} \Phi(t)\,dt - \int_{(-\infty,x]} \alpha(t)\,dt \int_{\mathbb{R}} \Phi(t)\,dt \right\} \text{ for } \Phi \in C_\infty(\mathbb{R}), \tag{1.56}$$

where $\alpha \in C_\infty(\mathbb{R})$ is real-valued and such that

$$\int_{\mathbb{R}} \alpha(t)\,dt = 1.$$

We see that

$$J: C_\infty(\mathbb{R}) \longrightarrow C_\infty(\mathbb{R}), \tag{1.57}$$

and by induction we have

$$D^k J^k \Phi = \Phi - \sum_{s=0}^{k-1} (D^s\alpha) \int_{\mathbb{R}} J^s\,\Phi(t)\,dt.$$

To determine $(C_\infty(\mathbb{R}))^\perp$ in $\mathcal{S}_k$ we need to find $f \in \mathcal{S}_k$ such that

$$((\mathcal{D}^*)^k\,\mathcal{D}^k\,\varphi\,,\,f) = 0 \text{ for all } \varphi \in C_\infty(\mathbb{R}).$$

Taking

$$\varphi = \gamma^{-1}\,J^k\,\gamma^2\,J^k\,\gamma^{-1}\,\Phi \text{ for a } \Phi \in C_\infty(\mathbb{R}), \tag{1.58}$$

we get

$$(\mathcal{D}^*)^k\,\mathcal{D}^k\,\varphi = \Phi - \sum_{s=0}^{k-1} \gamma\,(D^s\alpha) \int_{\mathbb{R}} J^s\,\gamma^{-1}(t)\,\Phi(t)\,dt +$$

$$- \sum_{s=0}^{k-1} \gamma\, D^k\, (\gamma^{-2}\, (D^s\alpha)) \int_{\mathbb{R}} J^s\, \gamma^2(t)\, J^k\, \gamma^{-1}(t)\, \Phi(t)\, dt.$$

From this we obtain that

$$(\Phi, f) = \sum_{s=0}^{k-1} (\gamma\, (D^s\alpha),\, f) \int_{\mathbb{R}} J^s\, \gamma^{-1}(t)\, \overline{\Phi(t)}\, dt +$$

$$+ \sum_{s=0}^{k-1} (\gamma\, D^k\, (\gamma^{-2}\, (D^s\alpha)),\, f) \int_{\mathbb{R}} J^s\, \gamma^2(t)\, J^k\, \gamma^{-1}(t)\, \overline{\Phi(t)}\, dt.$$

Now we shall again apply partial integration. To write this conveniently we introduce

$$J^* \varphi\, (x) = 2\pi i \{ \int_{[0,x]} \varphi(t)\, dt - (\alpha\, , \int_{[0,\bullet]} \varphi(t)\, dt) \},\ x \in \mathbb{R}. \tag{1.59}$$

We note that

$$(J^* \varphi, \psi) = (\varphi,\, J\, \psi) \text{ for all } \psi \in C_\infty(\mathbb{R}),\ \varphi \in C_\infty(\mathbb{R}).$$

Thus,

$$(\Phi, f) = \sum_{s=0}^{k-1} (-1)^s\, (\gamma\, (D^s\alpha),\, f)\, (\Phi,\, ((J^*)^s 1)\, \gamma^{-1}) +$$

$$+ \sum_{s=0}^{k-1} (-1)^{k+s}\, (\gamma\, D^k\, (\gamma^{-2}\, (D^s\alpha)),\, f)\, (\Phi,\, \gamma^{-1} (J^*)^k\, \gamma^2\, (J^*)^s 1).$$

Thus,

$$f = \{ \sum_{s=0}^{k-1} (-1)^s\, [(\gamma\, (D^s\alpha),\, f)\, ((J^*)^s 1) + (-1)^k\, (\gamma\, D^k\, (\gamma^{-2}\, (D^s\alpha)),\, f)\, (J^*)^k\, \gamma^2\, (J^*)^s 1] \}\, \gamma^{-1}. \tag{1.60}$$

By the polynomial character of the coefficient function as the argument goes to $+\infty$ and $-\infty$ and since

$$\gamma^{-1}(x) = e^{\pi x^2} \notin L_2(\mathbb{R}), \tag{1.61}$$

we shall obtain that $f \equiv 0$. First we observe that the first summation term in (1.60) is a polynomial of degree $(k-1)$ the second term —— let us call it T for now —— is of the form

$$T = \left(\int_{-\infty}^{x}\right)^k \gamma^2 Q \quad \text{modulo a polynomial of degree } (k-1),$$

where Q is a polynomial of degree (k-1) and $\left(\int_{-\infty}^{x}\right)^k$ means iterated evaluation of the anti-derivative operation $\int_{-\infty}^{x}$ that replaces J^* for convenience. By induction using partial integration it can be shown that

$$\left(\int_{-\infty}^{x}\right)^k \gamma^2 Q = \sum_{j=0}^{k-1} \frac{x^j}{j!} \int_{-\infty}^{x} \gamma^2(t)\, Q(t)\, \frac{(-t)^{k-1-j}}{(k-1-j)!}\, dt \, .$$

which in turn is equal to T modulo a polynomial of degree (k-1). Thus we have

$$f(x) = \left\{ \sum_{j=0}^{k-1} \frac{x^j}{j!} \int_{-\infty}^{x} \gamma^2(t)\, Q(t)\, \frac{(-t)^{k-1-j}}{(k-1-j)!}\, dt \;+ P(x) \right\} \gamma^{-1}(x),$$

for some polynomials P,Q of degree (k-1). Since the summation term tends to zero as $x \longrightarrow -\infty$ the condition $f \in L_2(\mathbb{R})$ implies that $P \equiv 0$. By the same reasoning applied to the limit $x \longrightarrow \infty$ we see that we must have

$$\sum_{j=0}^{k-1} \frac{x^j}{j!} \int_{-\infty}^{\infty} \gamma^2(t)\, Q(t)\, \frac{(-t)^{k-1-j}}{(k-1-j)!}\, dt = 0,$$

i.e.

$$\int_{-\infty}^{\infty} \gamma^2(t)\, Q(t)\, \frac{(-t)^j}{j!}\, dt = 0 \text{ for all } j \in \mathbb{N} \cup \{0\},\ j \le k-1.$$

Thus it follows that Q is also zero.

□

To obtain a better insight into the structure of the tempered Sobolev spaces $\mathcal{S}_k$, $k \in \mathbb{N}$, we show the following technical Lemma.

<u>Lemma 1.8</u>: *A function* f *is in* $\mathcal{S}_k$ *iff*

$$m^s D^t f \in L_2(\mathbb{R}) \text{ for } s,t \in \mathbb{N} \cup \{0\},\ s+t \le k\ ,\ k \in \mathbb{N}.$$

Moreover, the following estimates hold

$$\|m^s D^t f\|^2 \le \|f\|^2_{\mathscr{S},k} \le C \sum_{p+q\le k} \|m^p D^q f\|^2,$$

where C *is a generic constant depending only on* $k \in \mathbb{N}$.

Proof: The "if" part of the lemma is clear from the definition of $\mathscr{D}$ and Lemma 1.7. The "only if" part remains to be shown. The proof will be by induction with respect to k. The result is obviously true for k = 1 according to Lemma 1.6. Because of Lemma 1.7 we only have to show the assertion of the lemma for s + t = k + 1. Now, let $f \in \mathscr{S}_{k+1}$ then $\mathscr{D} f \in \mathscr{S}_k$ and by the induction hypothesis

$$m^s D^t \mathscr{D} f \in L_2(\mathbb{R}) \text{ for } s,t \in \mathbb{N} \cup \{0\},\ s + t \le k. \tag{1.62}$$

For $\Phi \in C_\infty(\mathbb{R})$ we have by the same calculation as in (1.42)

$$\begin{aligned}\|m^s D^t \mathscr{D} \Phi\|^2 &= \|m^s D^t (D+im)\Phi\|^2 = \|m^s D^{t+1} \Phi\|^2 + \|m^{s+1} D^t \Phi\|^2 + \\ &\quad + 2\,\mathrm{Re}\,(i\,(m^s D^{t+1} \Phi, m^{s+1} D^t \Phi)),\\ &= \|m^s D^{t+1} \Phi\|^2 + \|m^{s+1} D^t \Phi\|^2 + \\ &\quad - \frac{1}{\pi}\,\mathrm{Re}\,(\|m^s (D^t \Phi)', m^{s+1} D^t \Phi),\\ &= \|m^s D^{t+1} \Phi\|^2 + \|m^{s+1} D^t \Phi\|^2 - \frac{1}{\pi}\int_{\mathbb{R}} \frac{t^{2s+1}}{2} \frac{d}{dt} |D^t \Phi(t)|^2\,dt,\\ &= \|m^s D^{t+1} \Phi\|^2 + \|m^{s+1} D^t \Phi\|^2 + \frac{2s+1}{2\pi} \|m^s D^t \Phi\|^2.\end{aligned} \tag{1.63}$$

Since $s,t \in \mathbb{N} \cup \{0\}$ are arbitrary, (1.63) shows indeed not only that

$$m^s D^{t+1} f \in L_2(\mathbb{R}),\ m^{s+1} D^t f \in L_2(\mathbb{R}) \text{ for } s,t \in \mathbb{N} \cup \{0\},\ s + t = k,$$

which together with the induction hypothesis yields

$$m^s D^t f \in L_2(\mathbb{R}) \text{ for } s,t \in \mathbb{N} \cup \{0\},\ s + t = k + 1,$$

but also (by another induction argument) the estimates of the lemma. In fact, the constant C may be chosen (e.g.) as

$$C = \pi^{-k} \binom{k+7}{k} k!\ .$$

□

The importance of the spaces $\mathcal{S}_k$ lies in the fact that because of (1.41) we have

$$\mathcal{F}: \mathcal{S}_k \longrightarrow \mathcal{S}_k \text{ for } k \in \mathbb{N} \cup \{0\}, \tag{1.64}$$

and

$$\| \mathcal{F} \Phi \|_{\mathcal{S},k} = \| \Phi \|_{\mathcal{S},k}, \quad k \in \mathbb{N} \cup \{0\}. \tag{1.65}$$

Thus, $\mathcal{F}$ is a unitary mapping from $\mathcal{S}_k$ onto $\mathcal{S}_k$ for any $k \in \mathbb{N} \cup \{0\}$.

<u>Remark 1.6</u>: *The projective limit* $\mathcal{S}_\infty$ *of these tempered Sobolev spaces (i.e.* $\bigcap_{k\in\mathbb{N}} \mathcal{S}_k$ *with the induced topology) is the space of tempered Schwartz functions (compare e.g.* [4], *p. 175). It is interesting to note that* $\{\mathcal{S}_k\}_k$ *is indeed a compact chain of Hilbert spaces (for the terminology see* [7], chapter VI). *In fact, it follows from* (1.18) *and the estimate*

$$\sqrt{\int_{\mathbb{R}-I} |\Phi(x)|^2 \, dx} \le \frac{1}{L} \|\Phi\|_{\mathcal{S},1} \text{ for all } \Phi \in \mathcal{S}_1,$$

where $I = [-L,L]$, $L > 0$, *that*

$$\mathcal{S}_1 \hookrightarrow\hookrightarrow \mathcal{S}_0.$$

The compactness of the chain follows by induction.

We observe that

$$\mathcal{D} : \mathcal{S}_k \longrightarrow \mathcal{S}_{k-1},$$
$$\mathcal{D}^*: \mathcal{S}_k \longrightarrow \mathcal{S}_{k-1}, \tag{1.66}$$

where the latter is a consequence of Lemma 1.8. By the definition of the norm in $\mathcal{S}_k$, $k \in \mathbb{N} \cup \{0\}$, it is obvious that $\mathcal{D}$ is continuous as the mapping in (1.66) (in fact it is an isometry). The same continuity holds for $\mathcal{D}^*$, but this requires a more detailled consideration. We start with the observation that in generalization of (1.50), as is easily confirmed by induction with respect to s, we have

$$(\mathcal{D}^*)^k \mathcal{D}^s = \sum_{j=0}^{s} \binom{s}{j}\binom{k}{j} j!\, \pi^{-j} \mathcal{D}^{s-j} (\mathcal{D}^*)^{k-j},$$

$$\mathcal{D}^k (\mathcal{D}^*)^s = \sum_{j=0}^{s} (-1)^j \binom{s}{j}\binom{k}{j} j!\, \pi^{-j} (\mathcal{D}^*)^{s-j} \mathcal{D}^{k-j}, \quad s,k \in \mathbb{N},\ s \le k. \tag{1.67}$$

As a consequence of (1.67) (for s=k) we obtain a generalization of (1.52)

$$(\ \bullet\ ,\ \bullet\ )_{\mathcal{S},k} = \sum_{j=0}^{k} \binom{k}{k-j}^2 (k-j)!\, \pi^{j-k} ((\mathcal{D}^*)^j \bullet\ ,\ (\mathcal{D}^*)^j \bullet\ ),$$

$$((\mathcal{D}^*)^k \bullet\ ,\ (\mathcal{D}^*)^k \bullet\ ) = \sum_{j=0}^{k} (-1)^{k-j} \binom{k}{k-j}^2 (k-j)!\, \pi^{j-k} (\ \bullet\ ,\ \bullet\ )_{\mathcal{S},j}. \tag{1.68}$$

For s = 1 (1.67) yields

$$(\mathcal{D}^*)^k \mathcal{D} = \mathcal{D} (\mathcal{D}^*)^k + \frac{k}{\pi} (\mathcal{D}^*)^{k-1},$$

$$\mathcal{D}^k (\mathcal{D}^*) = \mathcal{D}^* \mathcal{D}^k - \frac{k}{\pi} \mathcal{D}^{k-1}. \tag{1.69}$$

This leads to

$$\|\mathcal{D}^* \bullet\|^2_{\mathcal{S},k} \le \|\bullet\|^2_{\mathcal{S},k+1} + \left(\frac{k}{\pi}\right)^2 \|\bullet\|^2_{\mathcal{S},k-1}, \quad k \in \mathbb{N}.$$

Since according to (1.42)

$$\|\bullet\|^2_{\mathcal{S},k} \le 2\pi \|\bullet\|^2_{\mathcal{S},k+1}, \quad k \in \mathbb{N} \cup \{0\}, \tag{1.70}$$

this implies

$$\|\mathcal{D}^* \bullet\|^2_{\mathcal{S},k} \le (1 + (2\ k)^2)\ \|\bullet\|^2_{\mathcal{S},k+1}, \quad k \in \mathbb{N} \cup \{0\}. \tag{1.71}$$

(The case k= 0 follows directly from (1.52)and is therefore included in (1.71)). This finally shows the continuity of $\mathcal{D}^*$.

To study the mapping properties of $\mathcal{D}$, $\mathcal{D}^*$ related to (1.66) we introduce the so-called tempered integration ([4], p.162 pp):

$$\mathcal{I}^* \Phi (x) = 2\pi i\ \gamma(x) \int_{(0,x]} \gamma^{-1} (t)\ \Phi(t)\ dt \text{ for all } \Phi \in C_\infty(\mathbb{R}). \tag{1.72}$$

$\mathcal{I}^*$ extends to a bounded, linear mapping

$$\mathcal{I}^* : \mathcal{S}_{k-1} \longrightarrow \mathcal{S}_k. \tag{1.73}$$

To see (1.73) we use the following lemma ([4], p. 167), Lemma 7.4.2).

Lemma 1.9: *The mapping* $\mathcal{F}^*$ *is continuous as a mapping from* $L_2(\mathbb{R})$ *into* $L_2(\mathbb{R})$.

Proof: It suffices to show

$$\|\mathcal{F}^* \Phi\| \leq C \|\Phi\| \text{ for all } \Phi \in C_\infty(\mathbb{R}), \tag{1.74}$$

where C is some real constant. First, we have

$$\begin{aligned} \|\mathcal{F}^* \Phi\|^2 &= 4\pi^2 \int_{\mathbb{R}} \gamma^2(x) \left| \int_{[0,x]} \gamma^{-1}(t)\, \Phi(t)\, dt \right|^2 dx \\ &= 4\pi^2 \int_{|x|<1} \gamma^2(x) \left| \int_{[0,x]} \gamma^{-1}(t)\, \Phi(t)\, dt \right|^2 dx + \\ &\quad + 4\pi^2 \int_{|x|\geq 1} \gamma^2(x) \left| \int_{[0,x]} \gamma^{-1}(t)\, \Phi(t)\, dt \right|^2 dx. \end{aligned} \tag{1.75}$$

We shall deal with the last two integrals in (1.75) separately. It is

$$\begin{aligned} 4\pi^2 \int_{|x|<1} \gamma^2(x) \left| \int_{[0,x]} \gamma^{-1}(t)\, \Phi(t)\, dt \right|^2 dx &\leq \\ &\leq 4\pi^2 \int_{|x|} \left(\int_{[0,x]} |\Phi(t)|\, dt \right)^2 dx, \end{aligned}$$

by Cauchy- Schwarz's inequality

$$\begin{aligned} &\leq 4\pi^2 \int_{|x|<1} |x| \left| \int_{[0,x]} |\Phi(t)|^2\, dt \right| dx \\ &\leq 4\pi^2 \int_{|x|<1} |x| \int_{\mathbb{R}} |\Phi(t)|^2\, dt\, dx \\ &\leq 4\pi^2 \int_{\mathbb{R}} |\Phi(t)|^2\, dt. \end{aligned} \tag{1.76}$$

The second integral can be estimated suitably by using partial integration and omitting negative boundary terms:

$$4\pi^2 \int_{|x|\geq 1} \gamma^2(x) \left| \int_{[0,x]} \gamma^{-1}(t)\, \Phi(t)\, dt \right|^2 dx \leq$$

$$\leq \pi \int_{|x|\geq 1} (4\pi\ |x|\ \gamma^2(x))\ \left| \int_{[0,x]} \gamma^{-1}(t)\ \Phi(t)\ dt\ \right|^2 dx$$

$$\leq \pi \int_{x\geq 0} (-\ \gamma^2(x))'\ \left| \int_{[0,x]} \gamma^{-1}(t)\ \Phi(t)\ dt\ \right|^2 dx\ +$$

$$+\ \pi \int_{x\leq 0} (\gamma^2(x))'\ \left| \int_{[0,x]} \gamma^{-1}(t)\ \Phi(t)\ dt\ \right|^2 dx\ ,$$

$$\leq 2\ \pi \int_{\mathbb{R}} \gamma^2(x)\ \left| \int_{[0,x]} \gamma^{-1}(t)\ \Phi(t)\ dt\ \right|\ \gamma^{-1}(x)\ |\Phi(x)|\ dx,$$

$$\leq 2\ \pi \int_{\mathbb{R}} \gamma(x)\ \left| \int_{[0,x]} \gamma^{-1}(t)\ \Phi(t)\ dt \right|\ |\Phi(x)|\ dx,$$

$$\leq 2\ \pi \int_{\mathbb{R}} \gamma(x)\ \left| \int_{[0,x]} \gamma^{-1}(t)\ \Phi(t)\ dt \right|\ |\Phi(x)|\ dx,$$

$$\leq \|\mathcal{J}^*\ \Phi\|\ \|\Phi\|\ . \qquad (1.77)$$

The latter estimate follows by another application of the Cauchy-Schwarz inequality. From (1.76), (1.77) we conclude that

$$\|\mathcal{J}^*\ \Phi\|^2 \leq 4\pi^2\ \|\Phi\|^2 + \|\mathcal{J}^*\ \Phi\|\ \|\Phi\| \text{ for all } \Phi \in C_\infty(\mathbb{R}). \qquad (1.78)$$

Inequality (1.78) is of the type

$$\alpha^2 \leq c\ \beta^2 + d\ \alpha\ \beta,\ \alpha,\beta,c,d \geq 0. \qquad (1.79)$$

This is equivalent to

$$(\alpha - \frac{d}{2}\ \beta)^2 \leq c\ \beta^2.$$

Now, if $\alpha \geq \frac{d}{2}\ \beta$ then this is eqivalent to

$$\alpha \leq (c^{1/2} + \frac{d}{2})\ \beta. \qquad (1.80)$$

If $\alpha \leq \frac{d}{2}\ \beta$ then (1.80) holds as well. Estimate (1.80) applied to (1.78) yields

$$\|\mathcal{J}^*\ \Phi\| \leq (2\pi + \frac{1}{2})\ \|\Phi\| \text{ for all } \Phi \in C_\infty(\mathbb{R}).$$

□

As a consequence of Lemma 1.9 we have

Lemma 1.10: *The mapping* $\mathcal{I}^*$ *is continuous, linear and one-to-one from* $\mathcal{S}_k$ *to* $\mathcal{S}_{k+1}$, $k \in \mathbb{N} \cup \{0\}$. *Moreover, we have*

$$\mathcal{D}^* \mathcal{I}^* \Phi = \Phi \text{ for } \Phi \in \mathcal{S}_k, \quad k \in \mathbb{N} \cup \{0\}. \tag{1.81}$$

Proof: The relationship (1.81) is obvious on $C_\infty(\mathbb{R})$. Linearity is also clear. It remains to show that $\mathcal{I}^*$ is continuous. Noting that for $\Phi \in C_\infty(\mathbb{R})$ we have

$$\mathcal{I}^* \Phi \in \mathcal{S}_k \text{ for any } k \in \mathbb{N} \cup \{0\},$$

we can derive

$$\|\mathcal{I}^* \Phi\|^2_{\mathcal{S},k+1} = \sum_{j=0}^{k+1} \binom{k+1}{k-j+1}^2 (k-j+1)!\, \pi^{j-k-1} ((\mathcal{D}^*)^j \mathcal{I}^*\Phi, (\mathcal{D}^*)^j \mathcal{I}^*\Phi),$$

$$= \sum_{j=1}^{k+1} \binom{k+1}{k-j+1}^2 (k-j+1)!\, \pi^{j-k-1} ((\mathcal{D}^*)^{j-1}\Phi, (\mathcal{D}^*)^{j-1}\Phi) +$$

$$+ (k+1)!\, \pi^{-k-1} (\mathcal{I}^*\Phi, \mathcal{I}^*\Phi)$$

$$= \sum_{j=1}^{k+1} \sum_{r=0}^{j-1} (-1)^{r-j+1} \binom{k+1}{k-j+1}^2 \binom{j-1}{j-r-1}^2 (k-j+1)!\, (j-r-1)!\, \pi^{r-k} \|\Phi\|^2_{\mathcal{S},r} +$$

$$+ (k+1)!\, \pi^{-k-1} (\mathcal{I}^*\Phi, \mathcal{I}^*\Phi), \tag{1.82}$$

$$\leq \Big\{ \sum_{j=1}^{k+1} \sum_{r=0}^{j-1} (-1)^{r-j+1} \binom{k+1}{k-j+1}^2 \binom{j-1}{j-r-1}^2 (k-j+1)!\, (j-r-1)!\, 2^{k-r} \Big\} \|\Phi\|^2_{\mathcal{S},k} +$$

$$+ (k+1)!\, \pi^{-k-1} \|\mathcal{I}^*\Phi\|_{\mathcal{S},0}.$$

Estimate (1.82) and Lemma 1.9 yield

$$\|\mathcal{I}^* \Phi\|^2_{\mathcal{S},k+1} \leq$$

$$\leq \Big\{ \sum_{j=1}^{k+1} \sum_{r=0}^{j-1} (-1)^{r-j+1} \binom{k+1}{k-j+1}^2 \binom{j-1}{j-r-1}^2 (k-j+1)!\, (j-r-1)!\, 2^{k-r} \Big\} \|\Phi\|^2_{\mathcal{S},k} +$$

$$+ \left(2 + \frac{1}{2\pi}\right) \pi^{-k+1} \|\Phi\|_{\mathcal{S},0}.$$

$$\leq \Big\{ \left(2 + \frac{1}{2\pi}\right) \pi\, 2^k +$$

$$+ \sum_{j=1}^{k+1} \sum_{r=0}^{j-1} (-1)^{r-j+1} \binom{k+1}{k-j+1}^2 \binom{j-1}{j-r-1}^2 (k-j+1)!\ (j-r-1)!\ 2^{k-r} \} \|\Phi\|^2_{\mathcal{S},k} .$$

This is the desired continuity estimate. By (1.81), (1.71) we also have

$$\|\Phi\|_{\mathcal{S},k} = \|\mathcal{D}^* \mathcal{I}^* \Phi\|^2_{\mathcal{S},k} \leq (1 + (2\ k)^2)\ \|\mathcal{I}^* \Phi\|^2_{\mathcal{S},k+1},\ k \in \mathbb{N} \cup \{0\}, \qquad (1.83)$$

which shows that $\mathcal{I}^*$ is indeed one-to-one.

□

Lemma 1.10 implies in particular that $\mathcal{D}^*$ as an operator from $\mathcal{S}_{k+1}$ to $\mathcal{S}_k$, $k \in \mathbb{N} \cup \{0\}$, is in fact onto, i.e.

$$\mathcal{S}_k = \mathcal{D}^* \mathcal{S}_{k+1}\ ,\ k \in \mathbb{N} \cup \{0\}. \qquad (1.84)$$

As one might have already noticed in the above there is a fundamental asymmetry between $\mathcal{D}$ and $\mathcal{D}^*$ which shows up in the fact that

$$\mathcal{S}_k \neq \mathcal{D}\ \mathcal{S}_{k+1},\ k \in \mathbb{N} \cup \{0\}. \qquad (1.85)$$

In fact, we have $\gamma \in N(\mathcal{D}^*)$, so that

$$\gamma \notin \mathcal{D}\ \mathcal{S}_1 .$$

Since $\gamma \in \mathcal{S}_k$

$$\mathcal{D}\ \mathcal{S}_{k+1} \subset \mathcal{D}\ \mathcal{S}_1,$$

for any $k \in \mathbb{N}$, (1.85) is obvious.

In order to calculate the ortho-complement of $\mathcal{D}\ \mathcal{S}_{k+1}$ in $\mathcal{S}_k$ we note that

$$\tilde{\mathcal{I}}\ \Phi\ (x) := 2\pi i\ \gamma^{-1}(x) \{ \int_{(-\infty,x]} \gamma(t)\ \Phi(t)\ dt\ +$$

$$- 2^{1/2} \int_{(-\infty,x]} \gamma^2(t)\ dt \int_{\mathbb{R}} \gamma(t)\ \Phi(t)\ dt \} \in \mathcal{S}_k \text{ for } \Phi \in C_\infty(\mathbb{R}),\ k \in \mathbb{N} \cup \{0\}. \qquad (1.86)$$

The operator $\tilde{\mathcal{I}}$ extends to a continuous linear mapping from $\mathcal{S}_k$ to $\mathcal{S}_{k+1}$, $k \in \mathbb{N} \cup \{0\}$ and defines in fact a pseudo-inverse of $\mathcal{D}$.

Lemma 1.11: *$\tilde{\mathcal{J}}$ is a continuous, linear mapping from $\mathcal{S}_k$ onto $\mathcal{S}_{k+1}$, $k \in \mathbb{N} \cup \{0\}$. Its pseudo-inverse is $\mathcal{D}$. We have, in particular*

$$\tilde{\mathcal{J}}\, \mathcal{D} = \mathrm{id}_k \text{ on } \mathcal{S}_k \text{ for } k \in \mathbb{N},$$

$$\mathcal{D}\, \tilde{\mathcal{J}} = \mathrm{id}_k - 2^{1/2}\, \gamma\, (\gamma, \bullet\,) \text{ on } \mathcal{S}_k \text{ for } k \in \mathbb{N} \cup \{0\}, \tag{1.87}$$

where id_k denotes the identity operator in $\mathcal{S}_k$.

Proof: Again, relationships (1.87) are clear on $C_\infty(\mathbb{R})$. It remains to establish continuity. To show this, is much more straight-forward than the corresponding theorem for $\mathcal{J}^*$. We have

$$\mathcal{D}^k\, \tilde{\mathcal{J}}\, \Phi = \mathcal{D}^{k-1}\Phi - 2^{1/2}\, (\mathcal{D}^{k-1}\, \gamma) \int_{\mathbb{R}} \gamma(t)\, \Phi(t)\, dt \text{ for } \Phi \in C_\infty(\mathbb{R}),\ k \in \mathbb{N}.$$

Thus,

$$\begin{aligned} \|\tilde{\mathcal{J}}\, \Phi\|_{\mathcal{S},k+1} &= \|\Phi - 2^{1/2}\gamma\, (\gamma, \Phi)\|_{\mathcal{S},k} \leq \\ &\leq \|\Phi\|_{\mathcal{S},k} + 2^{1/2}\, \|\gamma\|_{\mathcal{S},k} \left|\int_{\mathbb{R}} \gamma(t)\, \Phi(t)\, dt\right|, \\ &\leq \|\Phi\|_{\mathcal{S},k} + 2^{1/2}\, \|\gamma\|_{\mathcal{S},k}\, \|\gamma\|\, \|\Phi\|, \\ &\leq (1 + 2^{k+1/2}\, \pi^k \|\gamma\|_{\mathcal{S},k}\, \|\gamma\|)\, \|\Phi\|_{\mathcal{S},k}. \end{aligned}$$

□

Remark 1.7: *Note that $\rho_0 := 2^{1/2}\, \gamma\, (\gamma, \bullet\,)$ is in fact the orthogonal projection (with respect to the $L_2(\mathbb{R})$ inner product) on the null space $N(\mathcal{D}^*)$ of $\mathcal{D}^*$.*

For later purposes we note that the $L_2(\mathbb{R})$ adjoint $\mathcal{J}$ of $\mathcal{J}^*$ is given by

$$\mathcal{J}\, \Phi\, (x) = 2\pi i\, \gamma^{-1}(x)\, \left\{ \int_{(-\infty,x]} \gamma(t)\, \Phi(t)\, dt - h(x)\, (\gamma, \Phi) \right\} \tag{1.88}$$

for all $\Phi \in C_\infty(\mathbb{R})$, where h denotes the Heaviside function

$$h(t) \equiv \frac{1}{2}(1 + \mathrm{sgn}(t)) = \begin{cases} 1 \text{ if } t \geq 0, \\ \frac{1}{2} \text{ if } t = 0, \\ 0 \text{ if } t \leq 0. \end{cases}$$

We have from (1.81)

$$\mathcal{J}\,\mathcal{D} = \mathrm{id}_k \text{ on } \mathcal{S}_k\ ,\ k \in \mathbb{N}. \tag{1.89}$$

We have of course

$$\mathcal{J} \equiv \tilde{\mathcal{J}} \text{ on } \mathcal{D}\,\mathcal{S}_k\ ,\ k \in \mathbb{N}, \tag{1.90}$$

and thus trivially that $\mathcal{J}$ is continuous from $\mathcal{D}\,\mathcal{S}_k$ to $\mathcal{S}_k$.
Note, however, that

$$\mathcal{J}\,\gamma\ (x) = 2\pi i\ \mathrm{sgn}\ (x) \int_x^\infty \gamma^2(t)\ dt,$$

which because of the jump at 0 is only in $L_2(\mathbb{R})$. Thus, $\mathcal{J}$ does not extend to a continuous linear mapping from $\mathcal{S}_k$ to $\mathcal{S}_{k+1}$.

By an easy calculation we obtain the adjoint $\tilde{\mathcal{J}}^*$ of $\tilde{\mathcal{J}}$ as

$$\tilde{\mathcal{J}}^* = \mathcal{J}^* - 2^{1/2}\ \gamma\ (\gamma,\ \mathcal{J}^* \bullet\ ) \equiv (1-\rho_0)\ \mathcal{J}^*, \tag{1.91}$$

(compare Remark 1.7). This implies

$$\tilde{\mathcal{J}} = \mathcal{J}\ (1-\rho_0), \tag{1.92}$$

which can be verified directly by looking at (1.85). It will turn out that $\tilde{\mathcal{J}}$ and $\tilde{\mathcal{J}}^*$ are more useful with respect to generalization. Equality (1.97) together with the fact that $N(\mathcal{J}) = \{0\}$ yields

$$N(\tilde{\mathcal{J}}) = \mathrm{span}\ \{\gamma\}. \tag{1.93}$$

An immediate consequence of these considerations and the previous lemma is

<u>Lemma 1.12</u>: *For* $\mathcal{S}_k$ *the following orthogonal decomposition is valid*

$$\mathcal{S}_k = \mathcal{D}\,\mathcal{S}_{k+1} \oplus \mathrm{span}\ \{\tilde{\mathcal{J}}^k (\tilde{\mathcal{J}}^*)^k \gamma\},\ k \in \mathbb{N} \cup \{0\}. \tag{1.94}$$

<u>Proof</u>: Indeed, let $k \in \mathbb{N} \cup \{0\}$ be fixed and

$$(\mathcal{D}\,\Phi,\ f)_{\mathcal{S},k} = 0 \text{ for all } \Phi \in \mathcal{S}_{k+1}. \tag{1.95}$$

Choose

$$\Phi = \tilde{\mathcal{F}}\, \Psi, \quad \Psi \in C_\infty(\mathbb{R}),$$

then (1.95) implies

$$\begin{aligned} 2^{-1/2}\,(\Psi,\ f)_{\varphi,k} &= (\Psi,\ (\gamma,\ f)_{\varphi,k}\ \gamma), \\ &= (\Psi,\ (\gamma,f)_{\varphi,k}\ (\mathcal{D}^*)^k\ (\tilde{\mathcal{F}}^*)^k\ \gamma), \\ &= (\mathcal{D}^k\ \Psi,\ (\gamma,f)_{\varphi,k}\ (\tilde{\mathcal{F}}^*)^k\ \gamma), \\ &= (\mathcal{D}^k\ \Psi,\ (\gamma,f)_{\varphi,k}\ \mathcal{D}^k\ \tilde{\mathcal{F}}^k (\tilde{\mathcal{F}}^*)^k\ \gamma), \\ &= (\Psi,\ (\gamma,f)_{\varphi,k}\ \tilde{\mathcal{F}}^k (\tilde{\mathcal{F}}^*)^k\ \gamma)_{\varphi,k}\ , \end{aligned} \tag{1.96}$$

i.e.

$$f = 2^{1/2}\ (\gamma,f)_{\varphi,k}\ \tilde{\mathcal{F}}^k (\tilde{\mathcal{F}}^*)^k\ \gamma.$$

For (1.96) we made use of the fact that

$$\mathcal{D}^k\ \tilde{\mathcal{F}}^s = \mathcal{D}^{k-s} - 2^{1/2} \sum_{j=1}^{s} (\mathcal{D}^{k-j}\gamma)\ (\gamma,\ \tilde{\mathcal{F}}^{s-j}\ \bullet),\ s,k \in \mathbb{N} \cup \{0\}, s \le k. \tag{1.97}$$

Relationship (1.97) follows again by an induction argument. Noting that according to (1.67) and using Remark 1.4 we have

$$\mathcal{D}^s\ \gamma \in N((\mathcal{D}^*)^k) \text{ for } s < k, \tag{1.98}$$

equation (1.96) follows.

□

<u>Remark 1.8</u>: *Property (1.98) together with (1.68) implies in particular that the set of functions* $h_k = 2^{1/4}\ (k!)^{-1/2}\ \pi^{k/2}\ \mathcal{D}^k\ \gamma$, $k \in \mathbb{N} \cup \{0\}$, *is orthonormal in* $L_2(\mathbb{R})$. *The function* H_k *defined by*

$$H_k((2\pi)^{1/2}x) := (-i)^k\ 2^{k/2-1/4}\ (k!)^{1/2}\ \gamma^{-1}(x)\ h_k(x),$$

is a polynomial of degree k *known as* k*-th Hermite polynomial,* $k \in \mathbb{N} \cup \{0\}$, *(compare [23], chapter 5.6, [4], p. 141).*

The orthonormal set $\{h_k\}_{k \in \mathbb{N} \cup \{0\}}$ *is in fact complete in* $L_2(\mathbb{R})$ *(see [4], p. 140-141). Keeping Remark 1.4 in mind this shows that the spectrum* $\sigma(\mathcal{F})$ *of the unitary operator* $\mathcal{F}$ *is pure point spectrum* $P\sigma(\mathcal{F})$ *and we have*

$$\sigma(\mathcal{F}) = P\sigma(\mathcal{F}) = \{1,i,-1,-i\}.$$

For $\tilde{\mathcal{J}}^*$ we also note the adjoint formula to (1.97)

$$(\tilde{\mathcal{J}}^*)^s \, (\mathcal{D}^*)^k = (\mathcal{D}^*)^{k-s} - 2^{1/2} \sum_{j=1}^{s} ((\tilde{\mathcal{J}}^*)^{s-j}\gamma) \, (\gamma, \, (\mathcal{D}^*)^{k-j} \, \bullet), \qquad (1.99)$$

for $s,k \in \mathbb{N} \cup \{0\}, s \leq k$, as well as

$$\mathcal{D}^* \, \tilde{\mathcal{J}}^* = id_k \text{ on } \mathcal{S}_k \, , \; k \in \mathbb{N} \cup \{0\},$$

(as adjoint to (1.86)).

The purpose of studying "smaller" spaces lies in the fact that by duality we are able to extend many of the findings to "larger" spaces. To make this more precise we introduce tempered Sobolev spaces with negative indices. Let $\mathcal{S}_{-k}$ be the set of continuous linear functional (so-called distributions) on $\mathcal{S}_k$, $k \in \mathbb{N}$. The application of such a functional will be regarded as an extension of the $L_2(\mathbb{R})$ inner product and we thus introduce the notation

$$(f,\Phi) := f(\Phi) \text{ for } f \in \mathcal{S}_{-k} \, , \; \Phi \in \mathcal{S}_k, \; k \in \mathbb{N} \cup \{0\}. \qquad (1.100)$$

This is consistent with identifying the functional F corresponding to a function $f \in L_2(\mathbb{R})$ via

$$F(\Phi) \equiv (f,\Phi) \text{ for } \Phi \in C_\infty(\mathbb{R}). \qquad (1.101)$$

Note that by

$$(\alpha f + g)(\Phi) = \overline{\alpha} \, f(\Phi) + g(\Phi) \text{ for } f,g \in \mathcal{S}_{-k} \, , \; \Phi \in \mathcal{S}_k \, , \qquad (1.102)$$

$\mathcal{S}_{-k}$ becomes a linear space. By Riesz' representation theorem (see Appendix, Theorem A.3) we have a mapping

$$R_k \colon \mathcal{S}_{-k} \longrightarrow \mathcal{S}_k \qquad (1.103)$$

such that

$$f(\Phi) \equiv (f,\Phi) = (R_k f,\Phi)_{\mathcal{S},k}. \qquad (1.104)$$

Moreover,

$$\|f\|_{\mathcal{S},-k} := \sup_{\Phi \in \mathcal{S}_k - \{0\}} \frac{|f(\Phi)|}{\|\Phi\|_{\mathcal{S},k}} = \|R_k \, f\|_{\mathcal{S},k} \qquad (1.105)$$

for $f \in \mathcal{S}_{-k}$, $k \in \mathbb{N} \cup \{0\}$. Introducing the inner product

$$(\bullet \, , \, \bullet)_{\mathcal{S},-k} = (R_k \, \bullet \, , \, R_k \, \bullet \,)_{\mathcal{S},k} \, , \qquad (1.106)$$

$\mathcal{S}_{-k}$ becomes a Hilbert space and trivially R_k is unitary. In particular, R_k is onto since $(\Phi, \cdot)_{\mathcal{S},k}$ is a bounded linear functional on $\mathcal{S}_k$, $k \in \mathbb{N} \cup \{0\}$.

We have the following lemma.

Lemma 1.13: *The Hilbert space* $\mathcal{S}_{-k}$ *is continuously imbedded in* $\mathcal{S}_{-k-1}$, $k \in \mathbb{N} \cup \{0\}$.

Proof: Since $\mathcal{S}_{k+1} \subset \mathcal{S}_k$, a linear functional on $\mathcal{S}_k$ is also a linear functional on $\mathcal{S}_{k+1}$. Moreover, we have for $f \in \mathcal{S}_{-k}$ and all $\Phi \in \mathcal{S}_{k+1}$ using (1.42)

$$|f(\Phi)| \le C \, \|\Phi\|_{\mathcal{S},k} \le (2\pi)^{1/2} \, C \, \|\Phi\|_{\mathcal{S},k+1}.$$

This shows the lemma. □

Remark 1.9: (*i*) *The structure introduced by* (1.102) *looks different from the usual definition because of our choice identification in* (1.101). *The definition* (1.100) *has the advantage that the Riesz map* R_0 *becomes the identity on* $L_2(\mathbb{R})$ *(rather than conjugation) if elements of* $L_2(\mathbb{R})$ *are identified with functionals on* $L_2(\mathbb{R})$ *via* (1.101).

(*ii*) *The inductive limit* $\mathcal{S}_{-\infty}$ *of the tempered Sobolev spaces (i.e.* $\bigcup_{k \in \mathbb{N}} \mathcal{S}_k$ *equipped with the induced topology) is the space of tempered distributions (compare* [4] *p. 161 pp). In extension of Remark 1.6 we see that* $\{\mathcal{S}_k\}_k$ *remains a compact chain.*

We now shall extend the meaning of $\mathcal{D}$, $\mathcal{D}^*$, $\mathcal{F}$, $\mathcal{F}^*$ to these spaces. This can be done by letting

$$
\begin{aligned}
(\mathcal{D} f, \Phi) &= (f, \mathcal{D}^* \Phi) \text{ for } f \in \mathcal{S}_{-k} \text{ and all } \Phi \in \mathcal{S}_{k+1},\\
(\mathcal{D}^* f, \Phi) &= (f, \mathcal{D} \Phi) \text{ for } f \in \mathcal{S}_{-k} \text{ and all } \Phi \in \mathcal{S}_{k+1},\\
(\mathcal{F} f, \Phi) &= (f, \mathcal{F}^* \Phi) \text{ for } f \in \mathcal{S}_{-k} \text{ and all } \Phi \in \mathcal{S}_{k},\\
(\mathcal{F}^* f, \Phi) &= (f, \mathcal{F} \Phi) \text{ for } f \in \mathcal{S}_{-k} \text{ and all } \Phi \in \mathcal{S}_{k}.
\end{aligned}
\tag{1.107}
$$

We also extend the meaning of $\tilde{\mathcal{J}}^*$ and $\tilde{\mathcal{J}}$ to elements in $\mathcal{S}_{-k}$, $k \in \mathbb{N}$:

$$(\tilde{\mathcal{F}} f, \Phi) = (f, \tilde{\mathcal{F}}^* \Phi) \text{ for } f \in \mathcal{S}_{-k} \text{ and all } \Phi \in \mathcal{S}_k,$$

$$(\tilde{\mathcal{F}}^* f, \Phi) = (f, \tilde{\mathcal{F}} \Phi) \text{ for } f \in \mathcal{S}_{-k} \text{ and all } \Phi \in \mathcal{S}_k. \tag{1.108}$$

Similarly,

$$(\ell_{\alpha,\beta} f, \Phi) = (f, (\ell_{\alpha,\beta})^* \Phi) \text{ for } f \in \mathcal{S}_{-k} \text{ and all } \Phi \in \mathcal{S}_k,$$

$$(\rho_0 f, \Phi) = (f, \rho_0 \Phi) \text{ for } f \in \mathcal{S}_{-k} \text{ and all } \Phi \in \mathcal{S}_k \,,$$

$$(D f, \Phi) = (f, D \Phi) \text{ for } f \in \mathcal{S}_{-k} \text{ and all } \Phi \in \mathcal{S}_{k+1}, \tag{1.109}$$

$$(m f, \Phi) = (f, m \Phi) \text{ for } f \in \mathcal{S}_{-k} \text{ and all } \Phi \in \mathcal{S}_{k+1},\ k \in \mathbb{N} \cup \{0\}.$$

Note that

$$D\, \mathcal{S}_{k+1} \subset \mathcal{S}_k,\ m\, \mathcal{S}_{k+1} \subset \mathcal{S}_k, \tag{1.110}$$

as a consequence of Lemma 1.8.

As a first application we shall consider the functional

$$\begin{aligned} \delta : \mathcal{S}_1 &\longrightarrow \mathbb{C} \\ \Phi &\longrightarrow \Phi\Gamma_0 \end{aligned} \tag{1.111}$$

known as the Dirac δ distribution. Note that the trace $\Phi\Gamma_0$ (see (1.10)) is well-defined according to Lemma 1.2 ($\mathcal{S}_1 \subset H_1(\mathbb{R})$!). We shall show that

$$\delta \in \mathcal{S}_{-1}. \tag{1.112}$$

According to Lemma 1.11 we have

$$\Phi(0) = 2\pi i \, \{ \int_{(-\infty,0]} \gamma(t)\ \mathcal{D}\, \Phi(t)\ dt - \frac{1}{2} \int_{\mathbb{R}} \gamma(t)\ \mathcal{D}\, \Phi(t)\ dt \} \text{ for } \Phi \in C_\infty(\mathbb{R}).$$

Thus, using that $(a + b)^2 \leq 2(a^2 + b^2)$ and Schwarz's inequality we get the following Sobolev type estimate (compare Lemma 1.2)

$$\begin{aligned} |\Phi(0)| &\leq \pi \, | \int_{\mathbb{R}^-} \gamma(t)\ \mathcal{D}\, \Phi(t)\ dt | + | \int_{\mathbb{R}^+} \gamma(t)\ \mathcal{D}\, \Phi(t)\ dt |, \\ &\leq 2^{-1/4}\ \pi\ \|\Phi\|_{\mathcal{S},1}. \end{aligned} \tag{1.113}$$

This together with the density of $C_\infty(\mathbb{R})$ in $\mathcal{S}_1$ proves the desired continuity result that confirms (1.112). Thus we may apply the extended Fourier transform to δ, We have using the definitions

$$(\mathcal{F}\delta,\Phi) = (\delta,\ \mathcal{F}^*\Phi) = \int_{\mathbb{R}} \Phi(x)\ dx = (1,\ \Phi) \text{ for all } \Phi \text{ in } \mathcal{S}_1.$$

Thus,

$$\mathcal{F}\ \delta = 1 \in \mathcal{S}_{-1}\ . \tag{1.114}$$

We also have that

$$\mathcal{D}^*(2\pi i\ \gamma\ h) = \delta. \tag{1.115}$$

Indeed,

$$(2\pi i\ \gamma\ h\ ,\ \mathcal{D}\ \Phi) = -\int_{\mathbb{R}^+} \frac{d}{dt}\ (\gamma\ \Phi)(t)\ dt = \Phi(0) \text{ for } \Phi \in \mathcal{S}_1.$$

Note that $\gamma\ h \in L_2(\mathbb{R})$ (compare Lemma 1.13 below).

An immediate consequence of (1.113),(1.115) is

$$\mathcal{J}^*\ \mathcal{D}^* = id_k - \gamma\ \delta(\bullet)\ ,\ \mathcal{D}\ \mathcal{J} = id_k - (\gamma,\ \bullet\ )\ \delta \text{ on } \mathcal{S}_k,\ k \in \mathbb{N}. \tag{1.116}$$

That (1.116) is true on $C_\infty(\mathbb{R})$ can be seen by an easy calculation, but that it holds on $\mathcal{S}_k$ makes use of the continuity estimate (1.113) and Lemma 1.12. From (1.116) we obtain similar to (1.97)

$$(\mathcal{J}^*)^s\ (\mathcal{D}^*)^k = (\mathcal{D}^*)^{k-s} - \sum_{j=1}^{s} ((\mathcal{J}^*)^{k-j}\gamma)\ \delta((\mathcal{D}^*)^{s-j}\ \bullet\ ), \tag{1.117}$$

for $s,k \in \mathbb{N} \cup \{0\}, s \le k$.

As an application of (1.112) we now calculate the ranges

$$R(\mathcal{J}^*) \equiv \mathcal{J}^*\ \mathcal{S}_k \subset \mathcal{S}_{k+1}\ ,\ R(\tilde{\mathcal{J}}^*) \equiv \tilde{\mathcal{J}}^*\ \mathcal{S}_k \subset \mathcal{S}_{k+1}$$

of the mappings $\mathcal{J}^*$, $\tilde{\mathcal{J}}^*$.

<u>Lemma 1.14</u>: *The ranges of* $\mathcal{J}^*$ *and* $\tilde{\mathcal{J}}^*$ *can be characterized as the sets*

$$\mathcal{J}^*\ \mathcal{S}_k = \{\Phi \in \mathcal{S}_{k+1}\ |\ \Phi\upharpoonright_0 = 0\ \},$$

$$\tilde{\mathcal{J}}^*\ \mathcal{S}_k = \{\Phi \in \mathcal{S}_{k+1}\ |\ (\gamma,\ \Phi) = 0\} = (1-\rho_0)\ \mathcal{S}_{k+1},$$

for $k \in \mathbb{N} \cup \{0\}$.

<u>Proof</u>: Let $\Phi \in \mathcal{S}_{k+1}$ be arbitrary, $k \in \mathbb{N} \cup \{0\}$ fixed. Φ is in the range of $\mathcal{J}^*$ iff

$$\mathcal{J}^*\ \Psi = \Phi \text{ for some } \Psi \in \mathcal{S}_k.$$

By Lemma 1.10 this implies

$$\Psi = \mathcal{D}^* \Phi.$$

Application of $\mathcal{I}^*$ to both sides and using (1.116) yields

$$\Phi = \mathcal{I}^* \Psi = \Phi - \gamma \, \Phi\restriction_0 ,$$

or

$$\Phi\restriction_0 = 0.$$

On the other hand, if Φ satisfies this trace condition then

$$\mathcal{I}^* \mathcal{D}^* \Phi = \Phi.$$

This proves the first part of the lemma. As a consequence of this and (1.91) we obtain

$$\tilde{\mathcal{I}}^* \mathcal{S}_k = (1-\rho_0)\mathcal{I}^* \mathcal{S}_k = (1-\rho_0) \; \{\Phi \in \mathcal{S}_{k+1} \mid \Phi\restriction_0 = 0\}.$$

For an $f \in \tilde{\mathcal{I}}^* \mathcal{S}_k$ we must obviously have

$$(\gamma, f) = 0.$$

On the other hand, for any $f \in \mathcal{S}_{k+1}$ satisfying this orthogonality condition we can define

$$\Phi = f - \gamma \, f\restriction_0 \in \{\Phi \in \mathcal{S}_{k+1} \mid \Phi\restriction_0 = 0\} = \mathcal{I}^* \mathcal{S}_k ,$$

such that

$$f = (1-\rho_0) \; \Phi.$$

□

Remark 1.10: *This result shows that we cannot extend the tempered integration* $\mathcal{I}^*$ *in the same fashion as used in* (1.107)-(1.109). *The reason being that* $\mathcal{I}$ *is only defined as a continuous linear mapping from* $L_2(\mathbb{R})$ *to* $L_2(\mathbb{R})$. *Moreover, where it is defined in a sense suitable for our purposes, i.e. on* $\mathcal{D}\,\mathcal{S}_{k+1}$, $k \in \mathbb{N}$, *it coincides with* $\tilde{\mathcal{I}}$. *This is the background for a previous comment on our preference for* $\tilde{\mathcal{I}}$. *We note, however, that* $\mathcal{I}$ *could be extended, but we have no need to do so.*

Let us consider another example that will be of importance in a later chapter. We define the so-called comb function

$$Ш\,(\Phi) := \sum_{n\in\mathbb{Z}} \Phi(n),$$

for the moment only for $\Phi \in C_\infty(\mathbb{R})$. We shall see that Ш can be continuously extended such that

$$Ш \in \mathscr{S}_{-3}. \tag{1.118}$$

We first recall that according to Lemma 1.18 we have

$$\| m^s D^{k-s} \Phi \| \le \|\Phi\|_{\mathscr{S},k} \text{ for } \Phi \in \mathscr{S}_k,\ k \in \mathbb{N},\ s = 0,1,\ldots,k. \tag{1.119}$$

From (1.119) we obtain the following Sobolev type estimate ($s=k \in \mathbb{N}$)

$$|t^k \Phi(t)| \le 2\pi \left| \int_{[0,t]} D(x^k \Phi(x))\, dx \right|,$$

$$\le \left| \int_{[0,t]} |x^k D\Phi(x)|\, dx \right| + 2\pi \left| \int_{[0,t]} k\, |x^{k-1} \Phi(x)|\, dx \right|,$$

$$\le t^{1/2} \{ \| m^k D \Phi \| + 2\pi\, k\, \| m^{k-1} \Phi \| \},\ t \in \mathbb{R}.$$

Using (1.119) we get

$$|t^{k-1/2} \Phi(t)| \le (1+2\pi k)\, \|\Phi\|_{\mathscr{S},k+1} \text{ for all } \Phi \in \mathscr{S}_{k+1} \text{ and } t \in \mathbb{R}. \tag{1.120}$$

From (1.113), (1.120) we now see indeed that

$$|(Ш,\Psi)| \le \sup_{\Phi\in\mathscr{S}_3-\{0\}} \frac{|(Ш,\Phi)|}{\|\Phi\|_{\mathscr{S},3}} \|\Psi\|_{\mathscr{S},3},$$

$$= \sup_{\Phi\in\mathscr{S}_3-\{0\}} \frac{1}{\|\Phi\|_{\mathscr{S},3}} \left(\sum_{n\in\mathbb{Z}} \Phi(n) \right) \|\Psi\|_{\mathscr{S},3} \tag{1.121}$$

$$\le (1+4\pi) \left(1 + \sum_{n\in\mathbb{Z}-\{0\}} n^{-3/2}\right) \|\Psi\|_{\mathscr{S},3} \text{ for all } \Psi \in C_\infty(\mathbb{R}).$$

Since $C_\infty(\mathbb{R})$ is dense in $\mathscr{S}_3$, this continuity estimate finally proves (1.118).

With definitions (1.107)—(1.109) many of the properties studied above for tempered Sobolev spaces with positive index remain true for negative index. We shall collect some of the basic properties in the following lemmata.

As a first insight into the structure of the spaces we note

<u>Lemma 1.15</u>: *The set $C_\infty(\mathbb{R})$ is dense in $\mathscr{S}_{-k}$, $k \in \mathbb{N}$. The space $\mathscr{S}_{-k}$ can therefore be characterized as the completion of $C_\infty(\mathbb{R})$ with respect the norm $\|\bullet\|_{\mathscr{S},-k}$. Moreover, we have*

$$\mathscr{S}_{-k} = (\mathscr{D}^*)^k\, L_2(\mathbb{R}),\ k \in \mathbb{N} \cup \{0\}. \tag{1.122}$$

Proof: Let $f \in \mathscr{S}_{-k}$ and k be fixed. To see (1.122) we note that by definition we have

$$(f,\Phi) = (R_k\, f,\ \Phi)_{\mathscr{S},k} = (\mathscr{D}^k\, R_k\, f,\ \mathscr{D}^k\, \Phi)$$

for all $\Phi \in \mathscr{S}_k$. This says nothing else then

$$f = (\mathscr{D}^*)^k\, \mathscr{D}^k\, R_k\, f\ . \tag{1.123}$$

Since, $\mathscr{D}^k\, R_k\, f \in L_2(\mathbb{R})$, (1.122) is shown.

For the proof of the density result we first observe that

$$\begin{aligned}(\Phi,f)_{\mathscr{S},-k} &= (R_k\, \Phi,\ R_k f)_{\mathscr{S},k},\\ &= (\mathscr{D}^k\, R_k\, \Phi,\ \mathscr{D}^k\, R_k f) = ((\mathscr{D}^*)^k\, \mathscr{D}^k\, R_k\, \Phi,\ R_k f),\\ &= (\Phi,\ R_k f) \text{ for } \Phi \in C_\infty(\mathbb{R}).\end{aligned}$$

Thus, the ortho-complement of $C_\infty(\mathbb{R})$ in $\mathscr{S}_{-k}$ is in fact $\{0\}$. □

We can give an explicit representation for the Riesz map R_k , $k \in \mathbb{N} \cup \{0\}$. By (1.97), (1.98) and Lemma 1.10 we have

$$\begin{aligned}(\mathscr{D}^*)^k\, \mathscr{D}^k\, \tilde{\mathscr{I}}^k\, (\tilde{\mathscr{I}}^*)^k\Phi &= (\mathscr{D}^*)^k\, (\tilde{\mathscr{I}}^*)^k\Phi\ +\\ &\quad - 2^{1/2} \sum_{j=1}^{k} ((\mathscr{D}^*)^k\, \mathscr{D}^{k-j}\gamma)\ (\gamma,\ \tilde{\mathscr{I}}^{k-j}(\tilde{\mathscr{I}}^*)^k\Phi\ ),\\ &= (\mathscr{D}^*)^k\, (\tilde{\mathscr{I}}^*)^k\Phi,\\ &= \Phi \quad \text{for } \Phi \in L_2(\mathbb{R}),\ k \in \mathbb{N} \cup \{0\}.\end{aligned}$$

With $\Psi = \tilde{\mathscr{I}}^k\, (\tilde{\mathscr{I}}^*)^k\, \Phi$ this yields

$$\begin{aligned}(f,\ \tilde{\mathscr{I}}^k\, (\tilde{\mathscr{I}}^*)^k\Phi) \equiv (f,\Psi) &= (R_k f,\ \Psi)_{\mathscr{S},k},\\ &= (\mathscr{D}^k\, R_k\, f,\ \mathscr{D}^k\, \Psi),\\ &= (R_k f,\ (\mathscr{D}^*)^k\, \mathscr{D}^k\, \Psi) = (R_k f,\ \Phi)\end{aligned}$$

for any $\Phi \in L_2(\mathbb{R})$, $k \in \mathbb{N} \cup \{0\}$. In other words

$$R_k = (\tilde{\mathscr{I}}^k\, (\tilde{\mathscr{I}}^*)^k)^* = \tilde{\mathscr{I}}^k\, (\tilde{\mathscr{I}}^*)^k \text{ on } L_2(\mathbb{R}).$$

The Riesz map is thus $L_2(\mathbb{R})$-selfadjoint and invertible with inverse $(\mathscr{D}^*)^k\, \mathscr{D}^k$.
By definitions (1.108)

$$R_k = \tilde{\mathcal{J}}^k (\tilde{\mathcal{J}}^*)^k \text{ on } \mathcal{S}_{-k}, \; k \in \mathbb{N} \cup \{0\}. \tag{1.124}$$

With respect to the extension of $\mathcal{D}$, $\mathcal{D}^*$, $\tilde{\mathcal{J}}$, $\tilde{\mathcal{J}}^*$ to $\mathcal{S}_{-k}$, $k \in \mathbb{N}$, we have

<u>Lemma 1.16</u>: *The operators* $\mathcal{D}$, $\mathcal{D}^*$ *extend to continuous linear mappings from* $\mathcal{S}_{-k}$ *to* $\mathcal{S}_{-k-1}$, $k \in \mathbb{N} \cup \{0\}$. *The mappings* $\tilde{\mathcal{J}}$, $\tilde{\mathcal{J}}^*$ *extend accordingly to continuous linear maps from* $\mathcal{S}_{-k-1}$ *to* $\mathcal{S}_{-k}$, $k \in \mathbb{N} \cup \{0\}$. *In particular, we have*

$\mathcal{D}^* : \mathcal{S}_{-k} \longrightarrow \mathcal{S}_{-k-1}$ *is onto* $\mathcal{S}_{-k-1}$,

$\mathcal{D} : \mathcal{S}_{-k} \longrightarrow \mathcal{S}_{-k-1}$ *is an one-to-one,*

$\tilde{\mathcal{J}} : \mathcal{S}_{-k-1} \longrightarrow \mathcal{S}_{-k}$ *is onto* $\mathcal{S}_{-k}$,

$\tilde{\mathcal{J}}^* : \mathcal{S}_{-k-1} \longrightarrow \mathcal{S}_{-k}$ *is one-to-one ,*

for $k \in \mathbb{N} \cup \{0\}$. *Moreover, we have*

$$\tilde{\mathcal{J}} \mathcal{D} = \mathcal{D}^* \tilde{\mathcal{J}}^* = id_{-k} \text{ on } \mathcal{S}_{-k} \text{ for } k \in \mathbb{N} \cup \{0\},$$

$$\mathcal{D} \tilde{\mathcal{J}} = \tilde{\mathcal{J}}^* \mathcal{D}^* = id_{-k} - \rho_0 \quad \text{on } \mathcal{S}_{-k} \text{ for } k \in \mathbb{N},$$

We also have, that

$$N(\tilde{\mathcal{J}}) = \text{span } \{\gamma\},$$

and

$$\tilde{\mathcal{J}}^* \mathcal{S}_{-k-1} = \{f \in \mathcal{S}_{-k} \mid \rho_0 f = 0 \}.$$

<u>Proof</u>: Since $\mathcal{D}^* \tilde{\mathcal{J}}^* = \tilde{\mathcal{J}} \mathcal{D} = id_k$ on $\mathcal{S}_k$, it follows by definition that

$$\tilde{\mathcal{J}} \mathcal{D} = \mathcal{D}^* \tilde{\mathcal{J}}^* = id_{-k} \text{ on } \mathcal{S}_{-k}, \; k \in \mathbb{N}.$$

This shows that $\tilde{\mathcal{J}}$, $\mathcal{D}^*$ is onto. Since $\tilde{\mathcal{J}}$ and $\mathcal{D}^*$ are onto in spaces with positive index, we have again just by using the definition of the corresponding extensions. Similarly the other relations can be obtained. Continuity follows in the following way. Consider for example $\mathcal{D}^*$,

$$\|\mathcal{D}^* f\|_{\mathcal{S},-k} = \|R_k \mathcal{D}^* f\|_{\mathcal{S},k} = \sup_{\Phi \in \mathcal{S}_k - \{0\}} \frac{|(\mathcal{D}^* f, \Phi)|}{\|\Phi\|_{\mathcal{S},k}},$$

$$= \sup_{\Phi \in \mathcal{S}_k - \{0\}} \frac{|(f, \mathcal{D}\Phi)|}{\|\Phi\|_{\mathcal{S},k}},$$

$$= \sup_{\Phi \in \mathcal{S}_k - \{0\}} \frac{|(f, \mathcal{D}\,\Phi)|}{\|\mathcal{D}\Phi\|_{\mathcal{S},k-1}},$$

$$\leq \| f\|_{\mathcal{S},-k+1}.$$

Since all other continuity estimates can be obtained in the same fashion, we may leave the details to the reader. It remains to show that $N(\tilde{\mathcal{J}}) = \text{span}\,\{\gamma\}$. Now, let $f \in \mathcal{S}_{-k}$, $k \in \mathbb{N}$, be in $N(\tilde{\mathcal{J}})$, then

$$(f, \tilde{\mathcal{J}}^* \Phi) = 0 \text{ for all } \Phi \in \mathcal{S}_{k-1}.$$

Choosing $\Phi = \mathcal{D}^* \Psi$, $\Psi \in \mathcal{S}_k$, we get

$$(f, (1-\rho_0)\ \Psi) = 0 \text{ for all } \Psi \in \mathcal{S}_k,$$

or

$$(f,\Psi) = 2^{1/2}\ (f,\gamma)\ (\gamma.\Psi) \text{ for all } \Psi \in \mathcal{S}_k.$$

In other words, we have indeed

$$f = 2^{1/2}\ (f,\gamma)\ \gamma.$$

Since γ is certainly in $N(\tilde{\mathcal{J}})$ this proves the result.

Finally, $f \in \tilde{\mathcal{J}}^* \mathcal{S}_{-k-1}$ iff there is a $g \in \mathcal{S}_{-k-1}$ such that

$$(g, \tilde{\mathcal{J}}\,\Phi) = (f, \Phi).$$

Choosing $\Phi = \gamma$ shows indeed

$$(f,\gamma) = 0. \tag{1.125}$$

Reversely, if (1.125) is satisfied we have

$$\tilde{\mathcal{J}}^* \mathcal{D}^* f = f.$$

We note, in particular, that

$$\rho_0\ \tilde{\mathcal{J}}^* = 0. \tag{1.126}$$

□

To link up with a convergence concept used in [4], p. 169, ("tempered convergence") we show

<u>Lemma 1.17</u>: *A sequence* $\{g_n\}_n$ *converges to* g *in* $\mathcal{S}_{-k}$ *as* $n \longrightarrow \infty$ *iff there exists a sequence* $\{w_n\}_n$ *converging to a* w *in* $L_2(\mathbb{R})$ *such that* $g_n = (\mathcal{D}^*)^k w_n$ *and* $g = (\mathcal{D}^*)^k w$, $k \in \mathbb{N} \cup \{0\}$.

<u>Proof</u>: The "if" part of the lemma is clear from the proof of Lemma 1.15. Take

$$w_n := \mathcal{D}^k R_k g_n \quad \text{and } w := \mathcal{D}^k R_k g.$$

Convergence then follows, since (by (1.124), (1.126))

$$\mathcal{D}^k R_k = (\tilde{\mathcal{F}}^*)^k,$$

is continuous.

The "only if" part follows immediatly from estimating the k-th tempered derivative of the term $f \equiv w_n - w$.

$$\|g_n - g\|_{\mathcal{S},-k} = \|(\mathcal{D}^*)^k f\|_{\mathcal{S},-k} = \|R_k (\mathcal{D}^*)^k f\|_{\mathcal{S},k},$$

$$= \sup_{\Phi \in \mathcal{S}_k - \{0\}} \frac{|(f, \mathcal{D}^k \Phi)|}{\|\Phi\|_{\mathcal{S},k}},$$

$$= \sup_{\Phi \in \mathcal{S}_k - \{0\}} \frac{|(f, \mathcal{D}^k \Phi)|}{\|\mathcal{D}^k \Phi\|_{\mathcal{S},0}},$$

$$\leq \| f\|_{\mathcal{S},0} \equiv \|w_n - w\| .$$

□

For the Fourier transform $\mathcal{F}$ we obtain the following theorem which extends the properties previously stated to tempered Sobolev spaces $\mathcal{S}_k$ of arbitrary integer index k, i.e. $k \in \mathbb{Z}$.

<u>Theorem 1.4</u>: *The Fourier transform* $\mathcal{F}$ *is a unitary mapping from* $\mathcal{S}_k$ *onto* $\mathcal{S}_k$, $k \in \mathbb{Z}$. *In particular we have Parseval's equality*

$$\|\mathcal{F}^* \Phi\|_{\mathcal{S},k} = \|\mathcal{F} \Phi\|_{\mathcal{S},k} = \|\Phi\|_{\mathcal{S},k}, \quad \Phi \in \mathcal{S}_k, \quad k \in \mathbb{Z}.$$

Moreover, the Fourier transform enjoys the following properties:

(i) $\mathcal{F}^* D = -\mathcal{m}\, \mathcal{F}^*$, $\mathcal{F} D = \mathcal{m}\, \mathcal{F}$,

(ii) $D\, \mathcal{F} = -\mathcal{F}\, \mathcal{m}$, $D\, \mathcal{F}^* = \mathcal{F}^*\, \mathcal{m}$,

(iii) $i\, \mathcal{F}\, \mathcal{D} = \mathcal{D}\, \mathcal{F}$, $\mathcal{F}^*\, \mathcal{D} = i\, \mathcal{D}\, \mathcal{F}^*$,

(iv) $\mathcal{D}^* \mathcal{F}^* = i\, \mathcal{F}^* \mathcal{D}^*$, $i\, \mathcal{D}^* \mathcal{F} = \mathcal{F} \mathcal{D}^*$,

(v)
$$\ell_{\alpha,\beta}\, \mathcal{F} = \frac{1}{|\alpha|}\, \mathcal{F}\, e^{-2\pi i(\beta/\alpha)\mathbf{m}}\, \ell_{1/\alpha,0} = \frac{1}{|\alpha|}\, \mathcal{F}^*\, e^{2\pi i(\beta/\alpha)\mathbf{m}}\, \ell_{-1/\alpha,0},$$
$$\mathcal{F}^*\, \ell_{\alpha,\beta} = \frac{1}{|\alpha|}\, e^{-2\pi i(\beta/\alpha)\mathbf{m}}\, \ell_{1/\alpha,0}\, \mathcal{F}^*, \quad \mathcal{F}\, \ell_{\alpha,\beta} = \frac{1}{|\alpha|}\, e^{+2\pi i(\beta/\alpha)\mathbf{m}}\, \ell_{1/\alpha,0}\, \mathcal{F},$$

on $\mathcal{S}_k$, $k \in \mathbb{Z}$, $\alpha, \beta \in \mathbb{R}$, $\alpha \neq 0$.

In particular,

$$\ell_{-1,0}\, \mathcal{F} = \mathcal{F}\, \ell_{-1,0} = \mathcal{F}^*, \quad \ell_{-1,0}\, \mathcal{F}^* = \mathcal{F}^* \ell_{-1,0} = \mathcal{F},$$

$$\mathcal{F}\ \mathrm{sym} = \mathrm{Re}(\mathcal{F}) = \mathrm{Re}(\mathcal{F}^*) = \mathrm{sym}\ \mathcal{F},$$

$$\mathcal{F}\ \mathrm{asym} = i\ \mathrm{Im}(\mathcal{F}) = -i\ \mathrm{Im}(\mathcal{F}^*) = \mathrm{asym}\ \mathcal{F},$$

$$\mathcal{F}\, \mathcal{F} = \mathcal{F}^*\, \mathcal{F}^* = \ell_{-1,0},$$

$$\sigma_h\, \mathcal{F} = \mathcal{F}\, e^{+2\pi i h \mathbf{m}}, \quad \sigma_h\, \mathcal{F}^* = \mathcal{F}^*\, e^{-2\pi i h \mathbf{m}}$$

$$\mathcal{F}\, \sigma_h = e^{-2\pi i h \mathbf{m}}\, \mathcal{F}, \quad \mathcal{F}^*\, \sigma_h = e^{2\pi i h \mathbf{m}}\, \mathcal{F}^*,$$

on $\mathcal{S}_k$, $k \in \mathbb{Z}$, $h \in \mathbb{R}$.

Proof: We shall not prove all the statements of the theorem since the results follow in a rather straightforward way. In addition many of the formulas follow by taking adjoints from others. To give the flavour of the general procedure we shall demonstrate that $\mathcal{F}$ is an isometry and that the first equality of statement (v) is valid. That $\mathcal{F}$, $\mathcal{F}^*$ is unitary from $\mathcal{S}_k$ onto $\mathcal{S}_k$, $k \in \mathbb{N} \cup \{0\}$, is clear because of properties (1.41), (1.30), so we can assume that the index of the tempered Sobolev space to be considered is a negative integer —— say $-k$, where $k \in \mathbb{N}$. Let $k \in \mathbb{N}$ be fixed. It is

$$\|\mathcal{F} f\|_{\mathcal{S},-k} = \sup_{\Phi \in \mathcal{S}_k - \{0\}} \frac{|(f, \mathcal{F}^* \Phi)|}{\|\Phi\|_{\mathcal{S},k}}$$

$$= \sup_{\Phi \in \mathcal{S}_k - \{0\}} \frac{|(f, \mathcal{F}^* \Phi)|}{\|\mathcal{F}^* \Phi\|_{\mathcal{S},k}},$$

$$= \sup_{\Phi \in \mathscr{S}_k - \{0\}} \frac{|(f, \Phi)|}{\|\Phi\|_{\mathscr{S},k}} ,$$

since $\mathscr{F}: \mathscr{S}_k \longrightarrow \mathscr{S}_k$ is unitary. Thus indeed

$$\|\mathscr{F} f\|_{\mathscr{S},-k} = \|f\|_{\mathscr{S},-k} \text{ for } f \in \mathscr{S}_{-k},\ k \in \mathbb{N}.$$

To see

$$\mathscr{E}_{\alpha,\beta}\, \mathscr{F} = \frac{1}{|\alpha|}\, \mathscr{F}\, e^{-2\pi i (\beta/\alpha) \mathbf{m}}\, \mathscr{E}_{1/\alpha,0} \text{ on } \mathscr{S}_{-k} , \tag{1.127}$$

we note that according to (1.29) we have

$$\frac{1}{|\alpha|}\, \mathscr{F}^*\, \mathscr{E}_{1/\alpha,-\beta/\alpha} = e^{2\pi i \beta \mathbf{m}}\, \mathscr{E}_{\alpha,0}\, \mathscr{F}^* \text{ on } L_2(\mathbb{R}), \tag{1.128}$$

thus a-fortiori on $\mathscr{S}_k$. Now, noting that (1.127) is the $L_2(\mathbb{R})$ adjoint of (1.128), we just apply the definitions (1.107) — (1.109) to obtain (1.127).

□

Remark 1.11:

(i) *We notice that* sym $(1 \pm \mathscr{F})$ *remains a projection on the eigenspace* $N(\mathscr{F} \pm 1)$ *of* $\mathscr{F}$ *for the eigenvalue* ∓ 1 . *Orthogonality, however, is lost for* $k \neq 0$, *since* $\mathscr{D}$ *and* sym *do not commute.*

(ii) *By the above considerations the Fourier transform is extended to functions in* $L_1(\mathbb{R})$, *the set of absolutely integrable functions, as well. This is clear from the fact that* $L_1(\mathbb{R}) \subset \mathscr{S}_{-2}$. *In fact we have*

$$|(f,\Phi)| \leq 2\, \pi^{1/2} \int_{\mathbb{R}} |f(t)|\, dt\ \ \|\Phi\|_{\mathscr{S},2} \textit{ for all } f \in L_1(\mathbb{R}),\ \Phi \in \mathscr{S}_2.$$

For $f \in L_1(\mathbb{R})$ *the Fourier transform can be written as*

$$(\mathscr{F} f)(t) = \int_{\mathbb{R}} e^{-2\pi i t y}\, f(y)\, dy \textit{ for } f \in L_1(\mathbb{R})$$

here the integral is now an ordinary Lebesgue integral.

1.1.2.3 The Hartley transform.

In section 1.1.2 we developed a spectral representation $F_\sigma \equiv \sigma\, \mathscr{F}$ for the operator D. The choice of the coefficient σ , $|\sigma| = 1$, has of course little effect on what has been said about properties of the Fourier transform. However, since for a particular choice of this constant we can establish a relationship to the so-called Hartley transform that recently became well-known through the book by R.N. Bracewell, [9], we shall investigate the effect of different choices of σ. There is one set of relationships that apparently is affected by a different choice of σ. With regards to (1.32) we observe that

$$\operatorname{Re}(F_\sigma) \equiv \frac{F_\sigma + F_\sigma^*}{2} = \sigma \operatorname{Re}(\mathcal{F}) - i \operatorname{Im}(\sigma)\, \mathcal{F}^*,$$

$$\operatorname{Im}(F_\sigma) \equiv \frac{F_\sigma - F_\sigma^*}{2i} = \sigma \operatorname{Im}(\mathcal{F}) + \operatorname{Im}(\sigma)\, \mathcal{F}^*. \tag{1.129}$$

$\operatorname{Re}(F_\sigma)$, $\operatorname{Im}(F_\sigma)$ are commuting, selfadjoint operators on $\mathcal{S}_k$, $k \in \mathbb{Z}$. They satisfy the following equation:

$$\|F_\sigma \Phi\|^2_{\mathcal{S},k} = \|\operatorname{Re}(F_\sigma)\|^2_{\mathcal{S},k} + \|\operatorname{Im}(F_\sigma)\|^2_{\mathcal{S},k}. \tag{1.130}$$

For $\operatorname{Re}(F_\sigma)$, $\operatorname{Im}(F_\sigma)$ separately we have

<u>Lemma 1.18</u>: *For all $\Phi \in L_2(\mathbb{R})$ we have*

$$\|\operatorname{Re}(F_\sigma)\Phi\|^2 = |\operatorname{Re}(\sigma)|^2\ \|\operatorname{sym}(\Phi)\|^2 + |\operatorname{Im}(\sigma)|^2\ \|\operatorname{asym}(\Phi)\|^2,$$

$$\|\operatorname{Im}(F_\sigma)\Phi\|^2 = |\operatorname{Re}(\sigma)|^2\ \|\operatorname{asym}(\Phi)\|^2 + |\operatorname{Im}(\sigma)|^2\ \|\operatorname{sym}(\Phi)\|^2.$$

<u>Proof</u>: We calculate for $\Phi \in L_2(\mathbb{R})$

$$\begin{aligned}\|\operatorname{Re}(F_\sigma)\Phi\|^2 &= \|\sigma \operatorname{Re}(\mathcal{F})\Phi - i \operatorname{Im}(\sigma)\, \mathcal{F}^*\Phi\|^2 \\ &= |\sigma|^2\ \|\operatorname{Re}(\mathcal{F})\Phi\|^2 + |\operatorname{Im}(\sigma)|^2\ \|\mathcal{F}^*\Phi\|^2 + \\ &\qquad - 2 \operatorname{Re}\{i\ \sigma \operatorname{Im}(\sigma)\ (\operatorname{Re}(\mathcal{F})\Phi, \mathcal{F}^*\Phi)\}.\end{aligned}$$

Using (1.31)—(1.33) we get

$$\begin{aligned}\|\operatorname{Re}(F_\sigma)\Phi\|^2 &= |\sigma|^2\ \|\operatorname{Re}(\mathcal{F})\Phi\|^2 + |\operatorname{Im}(\sigma)|^2\ \|\mathcal{F}^*\Phi\|^2 + \\ &\qquad - 2 \operatorname{Re}\{i\ \sigma \operatorname{Im}(\sigma)\ (\operatorname{sym}(\Phi), \Phi)\}, \\ &= |\sigma|^2\ \|\operatorname{sym}(\Phi)\|^2 + |\operatorname{Im}(\sigma)|^2\ \|\Phi\|^2 - 2\ |\operatorname{Im}(\sigma)|^2\ \|\operatorname{sym}(\Phi)\|^2,\end{aligned}$$

$$= |\mathrm{Re}(\sigma)|^2 \; \|\mathrm{sym}(\Phi)\|^2 + |\mathrm{Im}(\sigma)|^2 \; \|\mathrm{asym}(\Phi)\|^2 .$$

A similar calculation shows the other equality. □

From this lemma it is apparent that for $|\mathrm{Im}(\sigma)| = |\mathrm{Re}(\sigma)| = 2^{1/2}$ we have that $2^{1/2}\,\mathrm{Re}(F_\sigma)$, $2^{1/2}\,\mathrm{Im}(F_\sigma)$ are isometries. In this case σ has to be equal to $\pm\, e^{\pm i\pi/4}$, where all signs are independent. Let us choose

$$\sigma = \sigma_0 \equiv e^{i\pi/4}, \tag{1.131}$$

and define

$$H := 2^{1/2}\, F_{\sigma_0} = 2^{1/2}\, e^{i\pi/4}\, \mathcal{F}, \tag{1.132}$$

$$H_+ := \mathrm{Re}(H),\; H_- := \mathrm{Im}(H), \tag{1.133}$$

then

$$H_\pm : L_2(\mathbb{R}) \longrightarrow L_2(\mathbb{R}), \text{ and } H_\pm \text{ is unitary.} \tag{1.134}$$

In fact, we have by another elementary calculation that $H_\pm$ is idem-potent, i.e.

$$H_\pm H_\pm = \mathrm{id}_0 \text{ on } L_2(\mathbb{R}). \tag{1.135}$$

As regards (1.129) we have

$$H_\pm = \mathrm{Re}(\mathcal{F}) \mp \mathrm{Im}(\mathcal{F}), \tag{1.136}$$

in particular this shows that H_+ , H_- are real transforms, i.e. transform real functions into real functions.

The mapping H_+ is known as the Hartley transform, see [9]. The relationship between $H_\pm$ and the Fourier transform can also be formulated using the projectors sym, asym. We have

$$\begin{aligned} H_\pm &= (\mathrm{sym} \pm i\ \mathrm{asym})\, \mathcal{F} , \\ &= \mathcal{F}\, (\mathrm{sym} \pm i\ \mathrm{asym}) = \mathcal{F}^*(\mathrm{sym} \mp i\ \mathrm{asym}) = (\mathrm{sym} \mp i\ \mathrm{asym})\, \mathcal{F}^*, \end{aligned} \tag{1.137}$$

and noting that (sym $\pm$ i asym) is $L_2(\mathbb{R})$-unitary with inverse (sym $\mp$ i asym)

$$\mathcal{F} = (\mathrm{sym} \mp i\ \mathrm{asym})\, H_\pm. \tag{1.138}$$

The caculation that proved the previous lemma does not work in $\mathcal{S}_k$ for $k \neq 0$. We have, however,

Lemma 1.19: *For the mappings* H_+ , H_- *we have that*

$$H_{\pm}: \mathcal{S}_k \longrightarrow \mathcal{S}_k \text{ is unitary, } k \in \mathbb{Z}.$$

Moreover, $H_{\pm}$ *satisfies the following commutation law with* $\mathfrak{D}$ *and* $\mathfrak{D}^*$,

$$H_{\pm}\, \mathfrak{D} = \pm\, \mathfrak{D}\, H_{\mp} \;,\; H_{\pm}\, \mathfrak{D}^* = \mp\, \mathfrak{D}^*\, H_{\mp} \text{ on } \mathcal{S}_k,\; k \in \mathbb{Z}. \tag{1.139}$$

Proof: That $H_{\pm}$ is unitary for $k \in \mathbb{N}$ is an immediate consequence of the first relationship (1.139) . They follow from the corresponding relationships for $\mathcal{F}$, $\mathcal{F}^*$:

$$\begin{aligned} 2\, H_{\pm}\, \mathfrak{D} &= \mathcal{F}\, \mathfrak{D} + \mathcal{F}^*\, \mathfrak{D} \mp i\, \mathcal{F}^*\, \mathfrak{D} \pm i\, \mathcal{F}\, \mathfrak{D}, \\ &= -i\, \mathfrak{D}\, \mathcal{F} + i\, \mathfrak{D}\, \mathcal{F}^* \pm \mathfrak{D}\, \mathcal{F}^* \pm \mathfrak{D}\, \mathcal{F}, \\ &= \pm\, \mathfrak{D}\, (\mp i\, \mathcal{F} \pm i\, \mathcal{F}^* + \mathcal{F}^* + \mathcal{F}), \\ &= \pm\, 2\, \mathfrak{D}\, H_{\mp} . \end{aligned}$$

The other relationship follows by taking $L_2(\mathbb{R})$-adjoints.

The unitarity negative index follows by a by now standard duality argument (compare the proof of Theorem 1.4). □

By (1.137), (1.138) properties known for the Fourier transform can be transferred to properties of the Hartley transform. We shall summarize some of the properties in the following theorem.

Theorem 1.5: *The transforms* H_+ , H_- *are idem-potent mappings on* $\mathcal{S}_k$, $k \in \mathbb{Z}$. *They are in close relationship to the Fourier transform as given in* formulae *(1.137) and (1.138).* H_+ *and* H_- *satisfy the following properties:*

(i) $H_{\pm}\, D = \pm\, i\, \mathbf{m}\, H_{\mp}$,

(ii) $D\, H_{\pm} = \mp\, i\, H_{\mp}\, \mathbf{m}$,

(iii) $H_{\pm}\, \mathfrak{D} = \pm\, \mathfrak{D}\, H_{\mp}$,

(iv) $H_{\pm}\, \mathfrak{D}^* = \mp\, \mathfrak{D}^*\, H_{\mp}$,

(v)

$$e_{\alpha,\beta}\, H_{\pm} = \frac{1}{|\alpha|} (H_{\pm} \cos(2\pi\, (\beta/\alpha)\, \mathbf{m}) \pm H_{\mp} \sin(2\pi(\beta/\alpha)\, \mathbf{m})\,)\, e_{1/\alpha,0}$$

$$H_{\pm}\, e_{\alpha,\beta} = \frac{1}{|\alpha|} (\cos(2\pi(\beta/\alpha)\mathbf{m})\, e_{1/\alpha,0}\, H_{\pm} \mp \sin(2\pi(\beta/\alpha)\mathbf{m})\, e_{1/\alpha,0}\, H_{\mp})\; ,$$

(vi) $$H_{\pm}\, e^{2\pi i(\beta/\alpha)\mathbf{m}} = |\alpha| \left\{ \frac{\mathcal{E}_{\alpha,\beta} + \mathcal{E}_{\alpha,-\beta}}{2} H_{+} \mp \frac{\mathcal{E}_{\alpha,\beta} - \mathcal{E}_{\alpha,-\beta}}{2} H_{-} \right\},$$

on $\mathcal{S}_k$, $k \in \mathbb{Z}$, $\alpha, \beta \in \mathbb{R}$, $\alpha \neq 0$.

In particular,

$$\mathcal{E}_{-1,0}\, H_{\pm} = H_{\pm}\, \mathcal{E}_{-1,0} = H_{\mp} \,,$$

$$H_{\pm}\, \text{sym} = \text{sym}\, H_{\pm} = \text{Re}(\mathcal{F}) = \text{sym}\, \mathcal{F} \,,$$

$$H_{\pm}\, \text{asym} = \text{asym}\, H_{\pm} = -\,\text{Im}(\mathcal{F}) = i\, \text{asym}\, \mathcal{F} \,,$$

$$\sigma_h\, H_{\pm} = H_{\pm} \cos(2\pi\, h\mathbf{m}) \mp H_{\mp} \sin(2\pi h\mathbf{m}) \,,$$

$$H_{\pm}\, \sigma_h = \cos(2\pi h\mathbf{m})\, H_{\pm} \pm \sin(2\pi h\mathbf{m})\, H_{\mp} \,,$$

$$H_{\pm}\, e^{2\pi i h\mathbf{m}} = \frac{\sigma_h + \sigma_{-h}}{2} H_{\pm} \pm i\, \frac{\sigma_h - \sigma_{-h}}{2} H_{\mp} \,,$$

on $\mathcal{S}_k$, $k \in \mathbb{Z}$, $h \in \mathbb{R}$.

<u>Proof</u>: Most of the properties have already been shown or are special cases of others. In fact, since (vi) follows from (v) by solving the linear system

$$\mathcal{E}_{\alpha,\beta}\, H_{\pm} = \frac{1}{|\alpha|} \left(H_{\pm} \cos(2\pi\, (\beta/\alpha)\, \mathbf{m}) \pm H_{\mp} \sin(2\pi(\beta/\alpha)\, \mathbf{m})\right) \mathcal{E}_{1/\alpha,0} \,,$$

$$\mathcal{E}_{\alpha,-\beta}\, H_{\pm} = \frac{1}{|\alpha|} \left(H_{\pm} \cos(2\pi\, (\beta/\alpha)\, \mathbf{m}) \mp H_{\mp} \sin(2\pi(\beta/\alpha)\, \mathbf{m})\right) \mathcal{E}_{1/\alpha,0} \,,$$

for $H_{\pm} \cos(2\pi\, (\beta/\alpha)\, \mathbf{m})$ and $H_{\mp} \sin(2\pi(\beta/\alpha)\, \mathbf{m})$, the only statement that remains to be shown is (v) and there only the first part, since the second part follows by taking $L_2(\mathbb{R})$-adjoints.

From (1.137) and (1.27) we have for α, $\beta \in \mathbb{R}$, $\alpha \neq 0$,

$$\begin{aligned}\mathcal{E}_{\alpha,\beta}\, H_{\pm} &= \frac{1}{|\alpha|}\, \mathcal{F}\, e^{-2\pi i(\beta/\alpha)\mathbf{m}}\, \mathcal{E}_{1/\alpha,0}\, (\text{sym} \pm i\, \text{asym}), \\ &= \frac{1}{|\alpha|}\, \mathcal{F}\, e^{-2\pi i(\beta/\alpha)\mathbf{m}}\, (\text{sym} \pm i\, \text{asym})\, \mathcal{E}_{1/\alpha,0} \,, \\ &= \frac{1}{|\alpha|}\, \mathcal{F}\, \{\cos(2\pi(\beta/\alpha)\mathbf{m})\, \text{sym} \pm i \cos(2\pi(\beta/\alpha)\mathbf{m})\, \text{asym} + \\ &\qquad -i \sin(2\pi(\beta/\alpha)\mathbf{m})\, \text{sym} \pm \sin(2\pi(\beta/\alpha)\mathbf{m})\, \text{asym}\}\, \mathcal{E}_{1/\alpha,0}\end{aligned}$$

$$= \frac{1}{|\alpha|} \mathcal{F} \{\text{sym} \cos(2\pi(\beta/\alpha)\boldsymbol{m}) \pm i \text{ asym} \cos(2\pi(\beta/\alpha)\boldsymbol{m}) +$$

$$-i \text{ asym} \sin(2\pi(\beta/\alpha)\boldsymbol{m}) \pm \text{sym} \sin(2\pi(\beta/\alpha)\boldsymbol{m})\} \, e_{1/\alpha,0}$$

$$= \frac{1}{|\alpha|} H_{\pm} \cos(2\pi(\beta/\alpha)\boldsymbol{m}) \pm H_{\mp} \sin(2\pi(\beta/\alpha)\boldsymbol{m}) \, e_{1/\alpha,0}.$$

The last five formulas are special cases of (v)and (vi).

□

1.1.3 Functions of D in the case of an unbounded interval.

The definition of functions of of an operator is one of the main usages of spectral representations. The function calculus provided by spectral theory allows to define normal operators that can be express as functions of the original selfadjoint operator. This applies in particular to the operator D and the Fourier tranform $\mathcal{F}$ as its spectral representation (as introduced in paragraph 1.1.2). The concept is purely spectral theoretical but it explains —— as we shall see ——the power of the Fourier transform in dealing with convolutions and differential operators having constant coefficients. In the light of the extension of $\mathcal{F}$ developed in section 1.1.2.2 it is apparent that the Fourier transform allows an extended function calculus that goes beyond what is provided by the spectral theorem. The tempered Sobolev space approach allows more general functions of D to be considered. The $L_2(\mathbb{R})$ setting, however, allows a more complete description of the domain of the operators produced by taking functions of **D**. We shall keep both aspects in mind.
Since $\mathcal{F}$ is a spectral representation for D, we have by definition

$$f(D) = \mathcal{F}^* f(\boldsymbol{m}) \mathcal{F}, \tag{1.139}$$

where (compare (1.25))

$$f(\boldsymbol{m}) \Phi (x) := f(x) \Phi(x),$$

for any Borel measurable function f defined a.e. w.r.t. Lebesgue measure. The domain D(f(D)) of f(D) can be characterized as

$$D(f(D)) = \{\Phi \in L_2(\mathbb{R}) \mid f(\boldsymbol{m}) \mathcal{F}(\Phi) \in L_2(\mathbb{R})\}. \tag{1.140}$$

1.1.3.1 Differential operators with constant coefficients.

The probably simplest case to consider is that f is a polynomial, say

$$p = \sum_{k=0}^{M} a_k m^k , \quad a_k \in \mathbb{C} , \quad k = 0, \ldots ,M , \quad a_M \neq 0 , \quad M \in \mathbb{N} , \text{ then}$$

$$p(D)\Phi = \sum_{k=0}^{M} a_k D^k \Phi \quad \text{for} \quad \Phi \in D(D^M),$$

and

$$\mathcal{F}(p(D)\Phi)(\lambda) = p(\lambda)\mathcal{F}(\Phi)(\lambda) \quad \text{for} \quad \Phi \in D(D^M) . \tag{1.141}$$

If p is a polynomial with real coefficients, the we can use the spectral theorem for D to obtain a spektral theorem for the selfadjoint operator p(D). As an illustration we shall consider in detail the case

$$p(\lambda) = \lambda^2 .$$

First, we note that $p(D) = D^2$ has spectrum $\sigma(D^2) = \overline{\mathbb{R}^+}$.

In order to obtain a spectral representation for D^2 , we observe the general fact that if $(\Pi(\lambda))_{\lambda\in\mathbb{R}}$ is the spectral family of D then

$$\pi(\lambda) := \begin{cases} \Pi(\sqrt{\lambda}) - \Pi(-\sqrt{\lambda}-0) & \text{if } \lambda \in \overline{\mathbb{R}^+} \\ 0 & \text{else} \end{cases} ,$$

is a spectral family for D^2 (see [2], p. 300, and Appendix, Theorem A.12). For $\lambda \geq 0$, $f \in L_2(\mathbb{R})$, we have

$$\|\pi(\lambda) f\|^2 = \|(\Pi(\sqrt{\lambda}) - \Pi(-\sqrt{\lambda}-0)) f\|^2 ,$$

$$= \int_{-\sqrt{\lambda}}^{\sqrt{\lambda}} |\mathcal{F}(f)(s)|^2 \, ds ,$$

$$= \frac{1}{2} \int_0^{\lambda} |\mathcal{F}(f)(\sqrt{t})|^2 \, t^{-1/2} \, dt + \frac{1}{2} \int_0^{\lambda} |\mathcal{F}(f)(-\sqrt{t})|^2 \, t^{-1/2} \, dt ,$$

$$= \int_0^{\lambda} \left| \frac{\mathcal{F}(f)(\sqrt{t}) - \mathcal{F}(f)(-\sqrt{t})}{2} \right|^2 t^{-1/2} \, dt +$$

$$+ \int_0^{\lambda} \left| \frac{\mathcal{F}(f)(\sqrt{t}) + \mathcal{F}(f)(-\sqrt{t})}{2} \right|^2 t^{-1/2} \, dt ,$$

$$= \int_0^{\lambda} |\mathcal{F}_{\sin}(f)(\sqrt{t})|^2 \, t^{-1/2} \, dt \; + \; \int_0^{\lambda} |\mathcal{F}_{\cos}(f)(\sqrt{t})|^2 \, t^{-1/2} \, dt \, , \qquad (1.141)$$

where we have set

$$\mathcal{F}_{\sin} := \frac{\mathcal{F} - \mathcal{F}^*}{2i} = \mathrm{Im}\,(\mathcal{F}) = -i \text{ asym } \mathcal{F},$$

$$\mathcal{F}_{\cos} := \frac{\mathcal{F} + \mathcal{F}^*}{2} = \mathrm{Re}(\mathcal{F}) = \text{sym } \mathcal{F}.$$

We have (in the sense of isometric $L_2(\mathbb{R})$ - extension of $\mathcal{F}_{\sin}$, $\mathcal{F}_{\cos}$ acting on functions with compact support)

$$\mathcal{F}_{\sin}(f)(s) = \int_{\mathbb{R}} \sin 2\pi \, s \cdot x \;\; f(x) \, dx \, , \qquad (1.142)$$

$$= 2 \int_{\mathbb{R}^+} \sin 2\pi \, s \cdot x \;\; \left(\frac{f(x) - f(-x)}{2} \right) dx \, , \qquad (1.143)$$

and similarly

$$\mathcal{F}_{\cos}(f)(s) = \int_{\mathbb{R}} \cos 2\pi \, s \cdot x \;\; f(x) \, dx \, , \qquad (1.144)$$

$$= 2 \int_{\mathbb{R}^+} \cos 2\pi \, s \cdot x \;\; \left(\frac{f(x) + f(-x)}{2} \right) dx \, . \qquad (1.145)$$

$\mathcal{F}_{\sin}$, $\mathcal{F}_{\cos}$ are called <u>Fourier sine</u> and <u>Fourier cosine</u> transform resp., because of equations (1.142), (1.144) , sometimes equipped with the attribute "two-sided". In this manner of speech (1.143) (or (1.145)) states, that the two-sided <u>Fourier-sine</u> (-<u>cosine</u>) <u>transform</u> of a function $f \in L_2(\mathbb{R})$ is equal to the one-sided <u>Fourier-sine</u> (-cosine) transform of the anti-symmetric (symmetric) part

$$\mathrm{asym}(f)(x) := \frac{f(x) - f(-x)}{2} \quad \left(\mathrm{sym}(f)(x) := \frac{f(x) + f(-x)}{2} \right)$$

of f . If we observe now that $\mathrm{asym}(L_2(\mathbb{R}))$ and $\mathrm{sym}(L_2(\mathbb{R}))$ are closed subspaces of $L_2(\mathbb{R})$, we see that

$$j_a : L_2(\mathbb{R}^+) \longrightarrow \mathrm{asym}(L_2(\mathbb{R}))$$

$$j_s : L_2(\mathbb{R}^+) \longrightarrow \operatorname{sym}(L_2(\mathbb{R}))$$

defined by

$$j_a(f)(x) := \begin{cases} f(x) & \text{if } x \in \overline{\mathbb{R}^+} \\ -f(-x) & \text{if } x \in \mathbb{R}^- \end{cases},$$

$$j_s(f)(x) := \begin{cases} f(x) & \text{if } x \in \overline{\mathbb{R}^+} \\ f(-x) & \text{if } x \in \mathbb{R}^- \end{cases},$$

(for all $x \in \mathbb{R}$ where f is defined), are unitary mappings. By the properties of the Fourier transform and (1.141) we have now that

$$V : L_2(\mathbb{R}) \longrightarrow L_2(\mathbb{R}^+) \oplus L_2(\mathbb{R}^+)$$

defined by

$$V(f)(\lambda) = j_s^* (\mathcal{F}_{\cos}(f))(\sqrt{\lambda})\lambda^{-1/4} \oplus j_a^* (\mathcal{F}_{\sin}(f))(\sqrt{\lambda})\lambda^{-1/4}, \qquad (1.146)$$

is a spectral representation for D^2. Since cosine, sine are even and odd functions respectively, application of j_s^*, j_a^* amount to restriction to $\mathbb{R}^+$. We summarize our results in

<u>Theorem 1.6</u>: D^2 *has spectrum* $\sigma(D^2) = \overline{\mathbb{R}^+}$ *of multiplicity* 2, *the spectrum is continuous and the unitary transformation* V *defined in* (1.146) *is a spectral representation for* D^2.

Considering the general case again we note that a polynomial funtion p(D) of degree M is a bounded mapping from $\mathcal{S}_k$ into $\mathcal{S}_{k-M}$. This follows for $k \geq M$ from Lemma 1.8. For k<M it can be concluded by the usual duality argument.

<u>1.1.3.2 Convolutions.</u>

More interesting than the case of a polynomial is of course to consider general functions of D. In the engineering literature application of f(D) is known under the term 'filtering'. Filtering with the operator corresponding to the Fourier transform $\mathcal{F} f$ of a function (or functional) f is known as convolution with f. To link up with the usual notations we define the convolution operation by

$$f * \Phi := (\mathcal{F}(f))(D)\ \Phi \text{ for } \Phi \in D((\mathcal{F}(f))(D)). \tag{1.147}$$

For well-behaved f, Φ the expression $f * \Phi$ is indeed a convolution integral. This is due to the fact that D commutes with the shift operator σ_h. Consequently σ_h commutes with functions of D:

$$\sigma_h\ f(D) = f(D)\ \sigma_h\ . \tag{1.148}$$

In fact, we see from Theorem 1.4 that σ_h is itself a function of D:

$$\sigma_h = e^{2\pi ihD},$$

or in convolution notation

$$\sigma_h\ f = (\sigma_h\delta) * f.$$

Remark 1.12: *In the light of this link between* D *and* σ_h *it is interesting to note that the symmetric difference quotient operator*

$$\frac{\sigma_h - \sigma_{-h}}{2ih} \equiv h^{-1}\ \mathrm{Im}(\sigma_h) = h^{-1}\ \sin(2\pi hD)$$

converges to $2\pi D$ *in the sense of strong resolvent convergence as* $h \longrightarrow 0$. *In fact, we have that*

$$f(h^{-1}\sin(2\pi hD)) \longrightarrow f(2\pi D) \text{ on } L_2(\mathbb{R}) \text{ as } h \longrightarrow 0,$$

for any bounded function f . *Moreover,*

$$h^{-1}\ \sin(2\pi hD) \longrightarrow 2\pi D \text{ on } H_1(\mathbb{R}) \text{ as } h \longrightarrow 0.$$

Now, if a function f(D) of the operator D happens to be an integral operator, i.e.

$$f(D)\ \Phi\ (x) = \int_{\mathbb{R}} \overline{K(x,y)}\ \Phi(y)\ dy, \tag{1.149}$$

where the kernel function K is defined on $\mathbb{R}^2$ such that — say, for simplicity — K is continuous and $\sup_{x\in\mathbb{R}} \int_{\mathbb{R}} |K(x,y)|\ dy$ is bounded, then we have from (1.148)

$$\int_{\mathbb{R}} K(x-h,y)\ \Phi(y)\ dy = \int_{\mathbb{R}} K(x,y)\ \Phi(y-h)\ dy = \int_{\mathbb{R}} K(x,y+h)\ \Phi(y)\ dy$$

for $\Phi \in C_\infty(\mathbb{R})$. In other words,

$$K(x-h,y) = K(x,y + h).$$

For $y = 0$ we obtain

$$K(x-h,0) = K(x,h).$$

Thus, (1.149) turns into

$$(f(D)\,\Phi)(x) = \int_{\mathbb{R}} K(x-y,0)\,\Phi(y)\,dy\,, x \in \mathbb{R}. \tag{1.150}$$

which is indeed a convolution integral. To justify definition (1.147) we have to demonstrate that the kernel function $K(\bullet,0)$ is in fact $\mathcal{F}^* f$. This can be seen from (1.150) by applying the Fourier transform and Fubini's theorem.

$$\begin{aligned} f(t)\,\mathcal{F}(\Phi)(t) &= \int_{\mathbb{R}} e^{-2\pi i t x} \int_{\mathbb{R}} K(x-y,0)\,\Phi(y)\,dy\,dx\,, \\ &= \int_{\mathbb{R}}\int_{\mathbb{R}} e^{-2\pi i t x}\,K(x-y,0)\,dx\,\Phi(y)\,dy\,, \\ &= \int_{\mathbb{R}}\int_{\mathbb{R}} e^{-2\pi i t x}\,K(x,0)\,dx\,e^{-2\pi i t y}\,\Phi(y)\,dy\,, \\ &= \mathcal{F}(K(\bullet,0))(t)\,\mathcal{F}\Phi(t). \end{aligned}$$

Thus,

$$K(\bullet,0) = \mathcal{F} f. \tag{1.151}$$

Of course, one may consider more general kernel functions, for our purposes this brief consideration is, however, sufficient, since we only want to justify that (1.147) is a generalization of the usual convolution integral. As far as our deductions go we see that we may take e.g the Gauss function γ or any $\Psi \in C_\infty(\mathbb{R})$ as kernel function, in which case we get smoothing operators

$$\gamma(D),\ \Psi(D)\ ,\ \Psi \in C_\infty(\mathbb{R}).$$

With definition (1.147) the convolution theorem becomes trivial. Obviously, we have

$$\mathcal{F}\,(f * \Phi) = \mathcal{F} f\, \mathcal{F}\,\Phi, \tag{1.152}$$

for $\Phi \in D(\mathcal{F}(f)(D)) = \{\Phi \in L_2(\mathbb{R}) \mid \mathcal{F}(f)(\boldsymbol{m})\,\mathcal{F}(\Phi) \in L_2(\mathbb{R})\}$. In view of the generalization to tempered Sobolev spaces we can extend the concept of functions of D — and thus convolutions — to those spaces. Apparently, if

$\overline{f(\boldsymbol{m})}: \mathcal{S}_j \longrightarrow \mathcal{S}_k$ is a bounded operator for some $j,k \in \mathbb{N} \cup \{0\}$, then a bounded operator from $\mathscr{S}_{-k}$ to $\mathscr{S}_{-j}$ is induced by the usual extension idea, i.e.

$$(f(\mathbf{m})\, g,\ \Phi) := (g,\ \overline{f(\mathbf{m})}\ \Phi) \text{ for } \Phi \in \mathscr{S}_j\ ,\ g \in \mathscr{S}_{-k}.$$

Then we define

$$f(D) = \mathscr{F}^*\ f(\mathbf{m})\ \mathscr{F},$$

and $f(D)$ is a bounded operator from $\mathscr{S}_{-k}$ into $\mathscr{S}_{-j}$ (as well as from $\mathscr{S}_j$ to $\mathscr{S}_k$); correspondingly we extend definition (1.147)

$$f * \Phi := \mathscr{F}(f)(D)\ \Phi \in \mathscr{S}_{-j} \text{ for } \Phi \in \mathscr{S}_{-k},\ k,j \in \mathbb{N} \cup \{0\}. \tag{1.153}$$

Relation (1.151) carries over accordingly. From this it is clear that in order to understand convolutions we only have to know about multiplication operators. In this connection we note

<u>Lemma 1.20</u>: *For $\mathscr{D}$ we have the following product rule*

$$\mathscr{D}^k\ (f\ g) = \sum_{j=0}^{k} \binom{k}{j}\ (D^j f)\ (\mathscr{D}^{k-j} g),$$

for $f \in \mathscr{S}_{k+2}$, $g \in \mathscr{S}_k$, $k \in \mathbb{N} \cup \{0\}$. *In particular, we have that the product* $(f\ g)$ *is in* $\mathscr{S}_k$.

<u>Proof</u>: First, we note that for $f \in \mathscr{S}_{k+2}$ we have by the same reasoning that led to (1.120)

$$|x^s\ D^t\ f(x)| \le 2\pi\ \left|\int_{[0,x]} D(u^s\ D^t f(u))\ du\right|,$$

$$\le \left|\int_{[0,t]} |u^s\ D^{t+1}\ f(u)|\ du\right| + 2\pi\ \left|\int_{[0,x]} s\ |u^{s-1}\ D^t f(u)|\ du\right|,$$

$$\le x^{1/2}\{\|\mathbf{m}^s D^{t+1}\ f\| + 2\pi\ s\ \|\mathbf{m}^{s-1}\ D^t\ f\|\},\ x \in \mathbb{R},\ s \ge 1.$$

Thus, using (1.119) we get

$$|x^{s-1/2}\ D^t f(x)| \le C\ \{\|\mathbf{m}\ D^{t+1}\ f\| + \|D^t\ f\|\} \le C\ \|f\|_{\mathscr{S},k+2}\ , \tag{1.154}$$

for all $f \in \mathscr{S}_{k+2}$ and $x \in \mathbb{R}$, where $(k + 1) \ge s+t$, $s \ge 1$. Here C denotes a generic constants depending only on $k \in \mathbb{N}$. The estimate (1.113) yields

$$|D^t f(0)| \le C\ \|D^t f\|_{\mathscr{S},1}$$

$$\leq C \{\|D^{t+1} f\| + \|\mathfrak{m}\, D^t f\|\} \leq C \|f\|_{\mathcal{S},k+1} , \qquad (1.155)$$

for $t \in \mathbb{N} \cup \{0\}$, $t \leq k$.

Estimates (1.154) (for s=1) and (1.155) imply that $D^t f$ is bounded for $t \leq k$. Thus it follows that $(f\, g) \in \mathcal{S}_k$. Moreover,

$$\mathfrak{D}(f\, g) = Df\, g + f\, \mathfrak{D}g,$$

and by induction the product rule holds.

□

Remark 1.13: *According to the definition of multiplication operators we have of course*

$$f\, g \equiv f(\mathfrak{m})\, g.$$

The product rule implies the following estimate .

Lemma 1.21: *We have the following estimates*

$$\|f(\mathfrak{m})\, g\|_{\mathcal{S},k} \leq C \sum_{j=0}^{k+1} \{\|\mathfrak{m}\, D^j f\| + \|D^j f\|\} \|g\|_{\mathcal{S},k} \leq C \|f\|_{\mathcal{S},k+2} \|g\|_{\mathcal{S},k} ,$$

$$\|\mathfrak{D}^k(f(\mathfrak{m})\, g) - f(\mathfrak{m})\, \mathfrak{D}^k g\| \leq C \Big(|(\delta, D f)| + \sum_{j=1}^{k} \{\|\mathfrak{m}\, D^{j+1} f\| + \|D^j f\|\}\Big) \|g\|_{\mathcal{S},k-1} ,$$

for $f \in \mathcal{S}_{k+2}$, $g \in \mathcal{S}_k$, $k \in \mathbb{N} \cup \{0\}$. *Here* C *is a generic constant depending only on* k.

Proof: The proof of the first inequality is immediate from (1.153). The second inequality is a refinement of the same reasoning. By Lemma 1.20 we have

$$\mathfrak{D}^k (f\, g) - f\, \mathfrak{D}^k\, g = \sum_{j=1}^{k} \binom{k}{j} (D^j f)\, (\mathfrak{D}^{k-j} g).$$

Estimates (1.154) then gives the result. Note that we preferred to leave the term $(\delta, D f) = (D f)(0)$ explicitely in the inequality, rather than using (1.155) to estimate it.

□

We now extend Lemma 1.21 to general indices.

<u>Lemma 1.22</u>: *The following estimate is valid*

$$\|f(\mathbf{m})\, g\|_{\mathcal{S},k} \le C \sum_{j=0}^{|k|+1} \{\|\mathbf{m}\, D^j\, f\| + \|D^j\, f\|\}\, \|g\|_{\mathcal{S},k} \le C\, \|f\|_{\mathcal{S},|k|+2}\, \|g\|_{\mathcal{S},k}\, ,$$

for $f \in \mathcal{S}_{|k|+2}$, $g \in \mathcal{S}_k$, $k \in \mathbb{Z}$. *Here* C *is a generic constant depending only on* k.

<u>Proof</u>: The result is stated in Lemma 1.20 for $k \in \mathbb{N} \cup \{0\}$. For negative index it follows by the usual duality argument. As a reminder we shall indicate the reasoning. Let $k \in \mathbb{N}$, then

$$\|f(\mathbf{m})\, g\|_{-k} = \sup_{\Phi\in\mathcal{S}_k-\{0\}} \frac{|(f(\mathbf{m})\, g, \Phi)|}{\|\Phi\|_{\mathcal{S},k}}\, ,$$

$$= \sup_{\Phi\in\mathcal{S}_k-\{0\}} \frac{|(g,\, \overline{f(\mathbf{m})}\Phi)|}{\|\Phi\|_{\mathcal{S},k}}\, ,$$

$$\le C\, \|g\|_{\mathcal{S},-k}\, C \sum_{j=0}^{k+1} \{\|\mathbf{m}\, D^j\, f\| + \|D^j\, f\|\}\, ,$$

$$\le C\, \|f\|_{\mathcal{S},k+2}\, \|g\|_{\mathcal{S},-k}.$$

□

As an application let us consider the family of smoothing operators $(\mathcal{F}\gamma_\alpha)(D)$, $\alpha \in \mathbb{R}$, $\alpha > 0$, generated by the functions

$$\gamma_\alpha(x) := \frac{1}{\alpha}\, \gamma(\frac{x}{\alpha}) = \frac{1}{\alpha}\, (\ell_{1/\alpha,0}\, \gamma)(x) \text{ for } \alpha > 0. \tag{1.156}$$

Theorem 1.4 yields (with Remark 1.4)

$$\mathcal{F}\, \gamma_\alpha = \ell_{\alpha,0}\, \gamma \text{ for } \alpha > 0. \tag{1.157}$$

Thus,

$$\mathcal{F}\gamma_\alpha\, (D) \equiv \gamma(\alpha\, D).$$

According to Lemma 1.8 we have that

$$\ell_{\alpha,0}\, \gamma : \mathcal{S}_k \longrightarrow \mathcal{S}_k\, , \; k \in \mathbb{N} \cup \{0\}.$$

By Lemma 1.21 we obtain (since $\gamma \in \mathcal{S}_j$ for any $j \in \mathbb{Z}$)

$$\|\gamma_\alpha * \Phi - \Phi\|_{\mathcal{S},k} = \|\mathcal{D}^k\, ((\ell_{\alpha,0}\, \gamma)\, \mathcal{F}\Phi) - \mathcal{D}^k\, \mathcal{F}\Phi\|,$$

$$\le \|\mathcal{D}^k\, ((\ell_{\alpha,0}\, \gamma)\, \mathcal{F}\Phi) - (\ell_{\alpha,0}\, \gamma)\, \mathcal{D}^k\, \mathcal{F}\Phi\| +$$

$$+ \| ((e_{\alpha,0}\,\gamma)-1)\,\mathcal{D}^k\,\mathcal{F}\Phi\|$$

$$\leq C\,(|(\delta,D\,\gamma)| + \sum_{j=1}^{k}\{\|m\,D^{j+1}\,(e_{\alpha,0}\,\gamma)\| + \|D^j\,(e_{\alpha,0}\,\gamma)\|\})\,\|\Phi\|_{\mathcal{S},k-1} +$$
$$+ \| ((e_{\alpha,0}\,\gamma)-1)\,\mathcal{D}^k\,\mathcal{F}\Phi\|,$$

for $\Phi \in \mathcal{S}_k$, $k \in \mathbb{N}$. Since

$$D\,(e_{\alpha,0}\,\Phi) = \alpha\,e_{\alpha,0}\,(D\,\Phi)\;,\;\Phi \in \mathcal{S}_k\;,\;k \in \mathbb{Z},$$

and

$$D\gamma(0) = 0,$$

we have

$$\|\gamma_\alpha * \Phi - \Phi\|_{\mathcal{S},k} \leq C\sum_{j=1}^{k}\alpha^j\,\{\alpha\,\|m\,(e_{\alpha,0}\,D^{j+1}\gamma)\| + \|e_{\alpha,0}(D^j\gamma)\|\}\,\|\Phi\|_{\mathcal{S},k-1} +$$
$$+ \| ((e_{\alpha,0}\,\gamma)-1)\,\mathcal{D}^k\,\mathcal{F}\Phi\|,$$
$$\leq C\sum_{j=1}^{k}\alpha^j\,\{\,\|e_{\alpha,0}\,(m\,D^{j+1}\,\gamma)\| + \|e_{\alpha,0}\,(D^j\,\gamma)\|\}\,\|\Phi\|_{\mathcal{S},k-1} +$$
$$+ \| ((e_{\alpha,0}\,\gamma)-1)\,\mathcal{D}^k\,\mathcal{F}\Phi\|,$$
$$\leq C\sum_{j=1}^{k}\alpha^{j-1/2}\,\{\|m\,D^{j+1}\,\gamma\| + \|D^j\,\gamma\|\}\,\|\Phi\|_{\mathcal{S},k-1} + \tag{1.158}$$
$$+ \| ((e_{\alpha,0}\,\gamma)-1)\,\mathcal{D}^k\,\mathcal{F}\Phi\|,$$

for $\Phi \in \mathcal{S}_k$, $k \in \mathbb{N}$. Since $((e_{\alpha,0}\,\gamma)-1)$ tends to zero point-wise, the last term vanishes as α approaches zero. Since the summation term obviously does the same, we have

$$\|\gamma_\alpha * \Phi - \Phi\|_{\mathcal{S},k} \longrightarrow 0 \text{ as } \alpha \longrightarrow 0+, \tag{1.159}$$

for any $\Phi \in \mathcal{S}_k$, $k \in \mathbb{N}$. For k=0 the summation term does not occur, so (1.159) remains true in this case, too. Moreover, we have

Theorem 1.7: *The smoothing operators $\gamma(\alpha\,D)$ converge to the identity on $\mathcal{S}_k$, i.e.*

$$\gamma(\alpha\,D)f \longrightarrow f \textit{ in } \mathcal{S}_k \textit{ as } \alpha \longrightarrow 0 \textit{ for any } f \in \mathcal{S}_k\;,\;k \in \mathbb{Z}.$$

Proof: The result has already been shown for $k \in \mathbb{N} \cup \{0\}$. For negative indices it follows by similar considerations as in the proof of Lemma 1.22

assuming that $f \in \mathcal{S}_k$ is such that $\mathcal{F}f \in \mathcal{S}_k$ has bounded support. Let $\Psi \in C_\infty(\mathbb{R})$, with $\Psi \equiv 1$ on support of f and $\Psi \equiv 0$ outside of a compact set K. Furthermore, let $k \in \mathbb{N}$ and $\alpha \in (0,1)$, then

$$(\gamma_\alpha * f - f, \Phi) = (((\mathcal{C}_{\alpha,0}\ \gamma) - 1)\ \mathcal{F}f,\ \mathcal{F}\Phi),$$
$$= (\ \mathcal{F}f,\ ((\mathcal{C}_{\alpha,0}\ \gamma) - 1)\ \mathcal{F}\Phi),$$
$$\leq C\ \|f\|_{\mathcal{S},-k}\ (\alpha^{1/2}\ \|\gamma\|_{\mathcal{S},k+2} + \|(((\mathcal{C}_{\alpha,0}\ \gamma)-1)\ \mathcal{D}^k(\Psi\ \mathcal{F}\Phi)\|),$$
$$\leq C\ \|f\|_{\mathcal{S},-k}\ (\alpha^{1/2}\ \|\gamma\|_{\mathcal{S},k+2} + \sup_{t\in K} |(\mathcal{C}_{\alpha,0}\ \gamma)(t)-1|\ \|\Phi\|_{\mathcal{S},k},$$

From this we obtain

$$\|\gamma_\alpha * f - f\|_{\mathcal{S},k} \leq C\ \|f\|_{\mathcal{S},-k}\ (\alpha^{1/2}\ \|\gamma\|_{\mathcal{S},k+2} + \sup_{t\in K} |(\mathcal{C}_{\alpha,0}\ \gamma)(t)-1|), \tag{1.60}$$

where the right-hand side apparently tends to zero as $\alpha \longrightarrow 0+$. From Lemma 1.15 we know that $C_\infty(\mathbb{R})$ is dense in $\mathcal{S}_{-k}$. So we shall show that the result follows in general by a density argument. We observe that by (1.158)

$$\|\gamma_\alpha * \Phi - \Phi\|_{\mathcal{S},k} \leq C \sum_{j=1}^{k} \alpha^{j-1/2}\ \{\|m\ D^{j+1}\ \gamma\| + \|D^j\ \gamma\|\}\ \|\Phi\|_{\mathcal{S},k-1} + \|\Phi\|_{\mathcal{S},k},$$

for all $\Phi \in \mathcal{S}_k$, $k \in \mathbb{N}$. Thus,

$$\|\gamma_\alpha * f - f\|_{\mathcal{S},-k} = \|(\mathcal{C}_{\alpha,0}\gamma - 1)\ \mathcal{F}\ f\|_{\mathcal{S},-k}\ ,$$
$$= \sup_{\Phi\in\mathcal{S}_k-\{0\}} \frac{|(\mathcal{C}_{\alpha,0}\gamma-1)\ \mathcal{F}f,\Phi)|}{\|\Phi\|_{\mathcal{S},k}}$$
$$= \sup_{\Phi\in\mathcal{S}_k-\{0\}} \frac{|(\mathcal{F}f,(\mathcal{C}_{\alpha,0}\gamma-1)\Phi)|}{\|\Phi\|_{\mathcal{S},k}},$$
$$\leq \|f\|_{\mathcal{S},-k}\ (C \sum_{j=1}^{k} \alpha^{j-1/2}\ \{\|m\ D^{j+1}\ \gamma\| + \|D^j\ \gamma\|\} + 1),$$
$$\leq C\ \|f\|_{\mathcal{S},-k}.$$

Here C is a generic constant only depending on $k \in \mathbb{N}$.

First choosing $\varphi \in C_\infty(\mathbb{R})$ such that $\|\gamma_\alpha * (\mathcal{F}^* \varphi - f) - (\mathcal{F}^* \varphi - f)\|$ is sufficiently small, then, using (1.160), α small so that $\|\gamma_\alpha * \mathcal{F}^* \varphi - \mathcal{F}^* \varphi\|$ is sufficiently small finally proves the convergence result via the triangle inequality.

□

Choosing $f = \delta$ in Theorem 1.7 shows in particular (note (1.114)) that

$$\gamma(\alpha D)\, \delta \equiv \gamma_\alpha * \delta = \gamma_\alpha \longrightarrow \delta \text{ in } \mathcal{S}_{-1}. \tag{1.161}$$

We shall call a family $(\delta_\alpha)_{\alpha \in A}$ of functions in $\mathcal{S}_k$, that has zero as an accumulation point of the index set $A \subset \mathbb{R}^+$ an $(\mathcal{S}_k, \mathcal{S}_{-j})$-$\delta$-family if

$$\delta_\alpha \longrightarrow \delta \text{ in } \mathcal{S}_{-j} \text{ as } \alpha \longrightarrow 0+.\ k \in \mathbb{N} \cup \{0\},\ j \in \mathbb{N}.$$

In particular,

$$\{\gamma_\alpha\}_\alpha \text{ is an } (\mathcal{S}_k, \mathcal{S}_{-j})\text{-}\delta\text{-family for any } k \in \mathbb{N} \cup \{0\} \text{ and } j \in \mathbb{N}. \tag{1.162}$$

To give another example of a δ-family consider the characteristic function Π of the unit interval $(-1/2,1/2)$

$$\begin{aligned} \Pi(t) &= (\sigma_{-1/2}\, h)(t) - (\sigma_{1/2}\, h)(t) \\ &= (\sigma_{-1/2}\, \mathrm{sgn})(t) - (\sigma_{1/2}\, \mathrm{sgn})(t), \\ &= \mathrm{sgn}(t + \tfrac{1}{2}) - \mathrm{sgn}(t - \tfrac{1}{2}). \end{aligned} \tag{1.163}$$

Obviously, $\Pi \in \mathcal{S}_0$ and its Fourier transform is a bounded, continuous function the so-called sinc function

$$\mathrm{sinc}(x) = \frac{\sin(\pi x)}{\pi x} \text{ for } x \neq 0,\ \mathrm{sinc}(0) = 1. \tag{1.164}$$

The sinc function has the remarkable property that it is zero on $\mathbb{Z}-\{0\}$. This can be employed for interpolation purposes as shall be discussed in section 1.1.3.3. Defining Π_α the rescaled version of Π analogous to γ_α by

$$\Pi_\alpha = \frac{1}{\alpha}\, e_{1/\alpha}\, \Pi \text{ for } \alpha > 0, \tag{1.165}$$

we have

$$\Pi_\alpha \longrightarrow \delta \text{ in } \mathcal{S}_{-1} \text{ as } \alpha \longrightarrow 0+. \tag{1.166}$$

Thus,

$$\{\Pi_\alpha\}_\alpha \text{ is an } (\mathcal{S}_0, \mathcal{S}_{-1})\text{-}\delta\text{-family.}$$

As a consequence for the operator family sinc(αD) we have, since all derivatives of sinc($\mathbf{m}$) are bounded, that

$$\text{sinc}(\alpha\, D) \longrightarrow \text{id}_k \text{ on } \mathcal{S}_k \text{ as } \alpha \longrightarrow 0+ \,,$$

for any $k \in \mathbb{Z}$.

Using the same pair of Fourier transforms we have

$$\Pi(\alpha\, D) \longrightarrow \text{id}_0 \text{ on } \mathcal{S}_0 \text{ as } \alpha \longrightarrow 0+,$$

and

$$\text{sinc}_{\alpha} := \frac{1}{\alpha}\, \ell_{\alpha.0}\, \text{sinc} \longrightarrow \delta \text{ in } \mathcal{S}_{-1} \text{ as } \alpha \longrightarrow 0+,$$

which shows that sinc_α is an $(\mathcal{S}_0, \mathcal{S}_{-1})$-$\delta$-sequence. We shall discuss an extension of $\Pi(\alpha\, D)$ to a larger class of functions in the next section.

1.1.3.2.1 Hilbert transform.

The sgn function is the asymmetric part of the Heaviside function h (the symmetric part is 1) and thus the knowledge of sgn(D) would automatically give us a description of h(D) via

$$h(D) = \frac{1}{2}\,(1 + \text{sgn}(D)). \tag{1.167}$$

From the spectral theorem we see that

sgn(D) is selfadjoint and unitary

(thus idem-potent).

$$\text{sgn}(D)\ \text{sgn}(D) = \text{id}_0 \text{ on } \mathcal{S}_0.$$

As before we have to find $\mathcal{F}^*$ sgn($\mathbf{m}$) in order to determine the convolution kernel for sgn(D). The simple estimate

$$\begin{aligned} |(\text{sgn}(\mathbf{m}),\Phi)| &\le \int_{\mathbb{R}} |\Phi(t)|\, dt, \\ &\le \left\{ \int_{\mathbb{R}} (1 + |t|^2)^{-1}\, dt \right\}^{1/2} \left\{ \int_{\mathbb{R}} (1 + |t|^2)\, |\Phi(t)|^2\, dt \right\}^{1/2}, \\ &\le 2^{3/2}\, \pi^{1/2}\, \|\Phi\|_{\mathcal{S},1}\,, \end{aligned} \tag{1.168}$$

shows that

$$\text{sgn}(\mathbf{m}) \in \mathcal{S}_{-1}.$$

Using the same technique as in the last section, we observe first that

$$(\ell_{\alpha,0}\, \Pi)\ \text{sgn}(\mathbf{m}) \in L_1(\mathbb{R}) \cap L_2(\mathbb{R}),\ \alpha \neq 0.$$

An analogous estimate to (1.168) shows that

$$(\ell_{\alpha,0}\,\Pi)\ \mathrm{sgn}(\mathbf{m}) \in \mathcal{S}_{-1},\quad \alpha \neq 0,$$

with a uniform bound with respect to α, and also that

$$(\ell_{\alpha,0}\,\Pi)\ \mathrm{sgn}(\mathbf{m}) \longrightarrow \mathrm{sgn}(\mathbf{m}) \text{ in } \mathcal{S}_{-1} \text{ as } \alpha \longrightarrow 0+\ , \tag{1.169}$$

We calculate

$$\begin{aligned}
\mathcal{F}^{*}((\ell_{\alpha,0}\,\Pi)\ \mathrm{sgn}(\mathbf{m}))(t) &= \int_{-1/\alpha}^{1/\alpha} e^{2\pi i t y}\,\mathrm{sgn}(y)\,dy\ ,\\
&= 2\,i \int_{0}^{1/\alpha} \sin(2\pi\,ty)\,dy\ ,\\
&= i\,\pi^{-1}\,t^{-1}\,(1-\cos(2\pi t/\alpha)) =: \varphi_{\alpha}(t),
\end{aligned}$$

for $\alpha > 0$. To study the limit if $\alpha \longrightarrow 0+$ we need a more detailled investigation. Note, however, that since the Fourier transform is unitary , we already know that this limit must be in $\mathcal{S}_{-1}$. For $\Phi \in \mathcal{S}_{1}$ and $\varepsilon \in (0,1)$ we have

$$\begin{aligned}
(\varphi_{\alpha}, \Phi) &= -i\,\pi^{-1}\int_{\mathbb{R}} t^{-1}\,(1-\cos(2\pi t/\alpha))\,\Phi(t)\,dt\ ,\\
&= -i\,\pi^{-1}\Big\{\Phi(0)\int_{-\varepsilon}^{\varepsilon} t^{-1}(1-\cos(2\pi t/\alpha))\,dt\ +
\end{aligned}$$

$$+\int_{-\varepsilon}^{\varepsilon} t^{-1}(1-\cos(2\pi t/\alpha))\,(\Phi(t)-\Phi(0))\,dt + \int_{|t|\geq\varepsilon} t^{-1}(1-\cos(2\pi t/\alpha))\,\Phi(t)\,dt\Big\},$$

$$\begin{aligned}
&=-i\pi^{-1}\Big\{\int_{-\varepsilon}^{\varepsilon} t^{-1}(1-\cos(2\pi t/\alpha))\,(\Phi(t)-\Phi(0))\,dt\ +\\
&\qquad + \int_{|t|\geq\varepsilon} t^{-1}(1-\cos(2\pi t/\alpha))\,\Phi(t)\,dt\Big\}.
\end{aligned}$$

The first integral in the brackets can be estimated according to Lemma 1.2:

$$\int_{-\varepsilon}^{\varepsilon} t^{-1}(1- \cos(2\pi t/\alpha)) \; (\Phi(t) - \Phi(0)) \, dt \le 8 \pi \varepsilon^{1/2} \, \| D \Phi \| . \qquad (1.170)$$

The second integral can be reformulated as

$$\int_{|t|\ge\varepsilon} t^{-1}(1- \cos(2\pi t/\alpha)) \, \Phi(t) \, dt = \int_{|t|\ge\varepsilon} t^{-1} \, \Phi(t) \, dt + - \int_{|t|\ge\varepsilon} \cos(2\pi t/\alpha)) \frac{\Phi(t)}{t} \, dt.$$

Now, the last integral in this expression can be estimated for $\Phi \in \mathscr{S}_2$ in the following way

$$\left| \int_{|t|\ge\varepsilon} \cos(2\pi t/\alpha)) \frac{\Phi(t)}{t} \, dt \right| \le \frac{\alpha}{2\pi} \left| \int_{|t|\ge\varepsilon} \sin(2\pi t/\alpha) \left(\frac{\Phi'(t)}{t} - \frac{\Phi(t)}{t^2} \right) dt \right| + + \alpha \, 2^{1/2} \, \varepsilon^{-1/2} \, \| \Phi \|_{\mathscr{S},1}$$

$$\le \alpha \, 2^{1/2} \, \varepsilon^{-3/2} \, \| \Phi \|_{\mathscr{S},1}. \qquad (1.71)$$

Thus, first choosing $\varepsilon \in (0,1)$ and then with ε fixed choosing $\alpha(\varepsilon) > 0$ we can arrange that $\alpha(\varepsilon) \longrightarrow 0+$ as $\varepsilon \longrightarrow 0+$ and

$$\varphi_{\alpha(\varepsilon)} - i \pi (1 - \ell_{\varepsilon,0}) \, \mathbf{m}^{-1} \longrightarrow 0 \text{ in } \mathscr{S}_{-1} \text{ as } \varepsilon \longrightarrow 0+.$$

This shows that

$$\mathscr{F}^* \operatorname{sgn}(\mathbf{m}) = i \pi^{-1} \lim_{\varepsilon \to 0+} (1 - \ell_{\varepsilon,0} \, \Pi) \, \mathbf{m}^{-1} , \qquad (1.172)$$

where the limit is taken in $\mathscr{S}_{-1}$. If we agree to understand $\mathbf{m}^{-1}$ in the sense of this limit (principle value) then we get

$$\mathscr{F}^* \operatorname{sgn}(\mathbf{m}) = i \pi^{-1} \mathbf{m}^{-1}, \; \mathscr{F} \operatorname{sgn}(\mathbf{m}) = - i \pi^{-1} \mathbf{m}^{-1}, \qquad (1.173)$$

and

$$\operatorname{sgn}(D) f = i \pi^{-1} \mathbf{m}^{-1} * f \text{ for } f \in \mathscr{S}_0. \qquad (1.174)$$

For the Heaviside function we obtain with (1.174)

$$h(D) f = \frac{1}{2} (f + \operatorname{sgn}(D)f) \text{ for } f \in \mathscr{S}_0. \qquad (1.175)$$

Of course we have corresponding to (1.173)

$$\mathscr{F}^* h = \frac{1}{2} (\delta + i \pi^{-1} \mathbf{m}^{-1}), \; \mathscr{F} h = \frac{1}{2} (\delta - i \pi^{-1} \mathbf{m}^{-1}). \qquad (1.176)$$

We shall now extend the Hilbert transform to $\mathscr{S}_{-k}$ spaces , $k \in \mathbb{N}$. For this we need several lemmas.

Let $\mathscr{S}_{k,0}$ denote the ortho-complement of $C_\infty(\mathbb{R}-\{0\})$ in $\mathscr{S}_k$, $k \in \mathbb{N}$, then we have the following description of this subspace.

Lemma 1.23: *For the ortho-complement* $\mathscr{S}_{k,0}$ *of* $C_\infty(\mathbb{R}-\{0\})$ *in* $\mathscr{S}_k$, $k \in \mathbb{N}$, *we have*

$$\mathscr{S}_{k,0} = \text{span}\left\{ \Phi \in \mathscr{S}_k \;\middle|\; \Phi = \sum_{j=0}^{k-1} \frac{x^j}{j!} \int_{-\infty}^{|x|} \gamma^2(t)\; Q_\pm(\mp t)\; \frac{(-t)^{k-1-j}}{(k-1-j)!}\, dt \right.$$

$$\left. \text{for } x \in \mathbb{R}^\mp,\; Q_\pm \text{ polynomials of degree } (k-1) \right\}.$$

The subspace $\mathscr{S}_{k,0}$ *is k-dimensional and its elements are (k-1)-times differentiable. The polynomials* Q_+ , Q_- *in the description of* $\mathscr{S}_{k,0}$ *satisfy the compatibility conditions*

$$\int_0^\infty \gamma^2(t)\; Q_-(-t)\; t^s\; dt + \int_0^\infty \gamma^2(t)\; Q_+(t)\; (-1)^s\; t^s\; dt = 0,$$

for all $s \in \mathbb{N} \cup \{0\}$, $s \leq k-1$.

Proof: Following the proof of Lemma 1.7 with an $\alpha \in C_\infty(\mathbb{R}^+)$ and again for an $\alpha \in C_\infty(\mathbb{R}^-)$ we can see that an element f in $(C_\infty(\mathbb{R}-\{0\}))^\perp$ must be of the form

$$f(x) = \sum_{j=0}^{k-1} \frac{x^j}{j!} \int_{-\infty}^{x} \gamma^2(t)\; Q_-(t)\; \frac{(-t)^{k-1-j}}{(k-1-j)!}\, dt \quad \gamma^{-1}(x) \quad \text{on } \mathbb{R}^-,$$

$$f(x) = -\sum_{j=0}^{k-1} \frac{x^j}{j!} \int_{x}^{\infty} \gamma^2(t)\; Q_+(t)\; \frac{(-t)^{k-1-j}}{(k-1-j)!}\, dt \quad \gamma^{-1}(x) \quad \text{on } \mathbb{R}^+,$$

where Q_+ , Q_- are polynomials of degree (k-1). In order to assure that $f \in \mathscr{S}_k$ we have to have compatibility conditions at zero satisfied. If Q_+ , Q_- are such that

$$\frac{d^s}{dx^s}\left\{\sum_{j=0}^{k-1} \frac{x^j}{j!}\left\{\int_{-\infty}^{x} \gamma^2(t)\, Q_-(t)\, \frac{(-t)^{k-1-j}}{(k-1-j)!}\, dt + \right.\right.$$

$$\left.\left. + \int_{x}^{\infty} \gamma^2(t)\, Q_+(t)\, \frac{(-t)^{k-1-j}}{(k-1-j)!}\, dt \right\}\right\}\Big|_0 = 0,$$

for $s \in \mathbb{N} \cup \{0\}$, $s \leq k-1$, then f can be extended continuously to an $f \in \mathcal{S}_k$. This set of conditions is a homogeneous linear system of k equations in 2k unknown coefficients. Ignoring all terms that have a positive power of x appearing in it we see that this is eqivalent to

$$\int_{-\infty}^{0} \gamma^2(t)\, Q_-(t)\, \frac{(-t)^{k-1-s}}{(k-1-s)!}\, dt + \int_{0}^{\infty} \gamma^2(t)\, Q_+(t)\, \frac{(-t)^{k-1-s}}{(k-1-s)!}\, dt = 0, \qquad (1.177)$$

or

$$\int_{0}^{\infty} \gamma^2(t)\, Q_-(-t)\, t^s\, dt + \int_{0}^{\infty} \gamma^2(t)\, Q_+(t)\, (-1)^s\, t^s\, dt = 0 \qquad (1.178)$$

for $s \in \mathbb{N} \cup \{0\}$, $s \leq k-1$. By this we have the following relationship between coefficients $c_j^{\pm}$ of $Q_{\pm}$:

$$(-1)^j\, c_j^-\int_{0}^{\infty} \gamma^2(t)\, t^{s+j}\, dt + (-1)^s\, c_j^+\int_{0}^{\infty} \gamma^2(t)\, t^{s+j}\, dt = 0\ ,$$

for $s,j \in \mathbb{N} \cup \{0\}$, $s,j \leq k-1$. Since the matrix

$$A = \left\{\int_{0}^{\infty} \gamma^2(t)\, t^{s+j}\, dt\right\}_{s,j},$$

is invertible we obtain the coefficients of Q_+ from those of Q_- and vice versa. In particular this shows that we have a k-dimensional solution set.

□

Next, we describe elements in $\mathcal{S}_{-k}$ that are zero on the closure $\mathcal{S}_{k,0}^{\perp}$ of $C_\infty(\mathbb{R}-\{0\})$ in $\mathcal{S}_k$, $k \in \mathbb{N}$. Such an element is said to have support $\{0\}$. The support supp(f) of an element $f \in \mathcal{S}_k$ is defined to be the complement of the largest open set K for which

$$(f,\Phi) = 0 \text{ for all } \Phi \in C_\infty(K),\ k \in \mathbb{Z}.$$

For sake of easy reference we introduce the notation

$$\mathscr{S}_{-k,0} := \{f \in \mathscr{S}_{-k} \mid \operatorname{supp}(f) = \{0\}\},\ k \in \mathbb{N}.$$

<u>Lemma 1.24</u>: *For* $\mathscr{S}_{-k,0}$ *we have*

$$\mathscr{S}_{-k,0} = \operatorname{span}\{D^s\,\delta\}_s,$$

where s ranges over all non-negative integers less than k, k$\in \mathbb{N}$.

<u>Proof</u>: Let $f \in \mathscr{S}_{-k}$ be such that

$$(f,\Phi) = 0 \text{ for all } \Phi \in C_\infty(\mathbb{R}-\{0\})\ k\in \mathbb{N}.$$

If $\alpha \in C_\infty(\mathbb{R})$ such that $\alpha \equiv 1$ in a neighbourhood of zero then

$$(1 - (e_{\lambda,0}\alpha))\,\Phi \in C_\infty(\mathbb{R}-\{0\}) \text{ for any } \Phi \in C_\infty(\mathbb{R}),\ \lambda \in \mathbb{R}^+.$$

Thus,

$$(f,\Phi) = (f,(e_{\lambda,0}\alpha)\,\Phi) \quad \text{for } \Phi \in C_\infty(\mathbb{R}). \tag{1.179}$$

Because of the cut-off function $(e_{\lambda,0}\alpha)$, we may write this as

$$(f,\Phi) = (f,(e_{\lambda,0}\alpha)\,\{\Phi - T_k\}) + (f,(e_{\lambda,0}\alpha)\,T_k),$$

where

$$T_k = \alpha \sum_{s=0}^{k-1} \gamma^{-1}\,(\mathscr{D}^s\,\Phi)(0)\,\frac{(2\pi i x)^s}{s!}.$$

Noting that

$$(\mathscr{D}^s\,\{\Phi - T_k\})(0) = 0 \text{ for } s \in \mathbb{N}\cup\{0\},\ s \le k-1,$$

we have

$$|\mathscr{D}^s\,\{\Phi - T_k\})(x)| \le C\,|x|^{k-s},\ s \in \mathbb{N}\cup\{0\},\ s \le k. \tag{1.180}$$

Moreover, by applying the product rule

$$|(f,(e_{\lambda,0}\alpha)\,\{\Phi - T_k\})| \le \|f\|_{\mathscr{S},-k}\,\|(e_{\lambda,0}\alpha)\,\{\Phi - T_k\}\|_{\mathscr{S},k},$$

$$\le \|f\|_{\mathscr{S},-k}\,\|\mathscr{D}^k(e_{\lambda,0}\alpha)\,\{\Phi - T_k\}\|,$$

$$\le C \sum_{j=0}^{k} \|f\|_{\mathscr{S},-k}\,\|D^j(e_{\lambda,0}\alpha)\,\mathscr{D}^{k-j}\{\Phi - T_k\}\|.$$

Using estimate (1.180) we get

$$|(f, (e_{\lambda,0}\alpha)\{\Phi - T_k\})| \le C \sum_{j=0}^{k} \lambda^{-1/2} \|f\|_{\mathcal{S},-k},$$

for some constant C independent of $\lambda > 0$. We therefore get

$$|(f, (e_{\lambda,0}\alpha)\{\Phi - T_k\})| \longrightarrow 0 \text{ as } \lambda \longrightarrow \infty,$$

and so, by (1.179),

$$(f, \Phi) = \lim_{\lambda\to\infty} (f, (e_{\lambda,0}\alpha) T_k) = (f, T_k). \tag{1.181}$$

Using the definition of T_k we obtain from (1.181)

$$(f,\Phi) = \sum_{s=0}^{k-1} (f, \frac{(2\pi i \mathfrak{m})^s}{s!} \alpha \gamma^{-1}) (\mathcal{D}^s \Phi)(0), \tag{1.182}$$

$$= \sum_{s=0}^{k-1} \frac{(2\pi i)^s}{s!} (f, \mathfrak{m}^s \alpha \gamma^{-1}) (\delta, \mathcal{D}^s \Phi),$$

or

$$f = \sum_{s=0}^{k-1} \frac{(2\pi i)^s}{s!} (f, \mathfrak{m}^s \alpha \gamma^{-1}) (\mathcal{D}^*)^s \delta.$$

The result follows, since

$$(\mathcal{D}^*)^s \delta \equiv D^s \delta. \tag{1.183}$$

□

A result that links up the statement in Lemma 1.23 and Lemma 1.24 is

Lemma 1.25: *The subspace* $\mathcal{S}_{-k,0}$, $k \in \mathbb{N}$, *can be described by*

$$\mathcal{S}_{-k,0} = (\mathcal{D}^*)^k \mathcal{D}^k \mathcal{S}_{k,0}.$$

Proof: If f has support {0} then

$$(f, \Phi) = 0 \text{ for all } \Phi \in C_\infty(\mathbb{R}-\{0\}).$$

This is the same as saying

$$R_k f \in \mathcal{S}_{k,0}.$$

Since $(\mathcal{D}^*)^k \mathcal{D}^k$ is the inverse of R_k we have

$$f = (\mathcal{D}^*)^k \mathcal{D}^k R_k f.$$

On the other hand any $f \in \mathscr{S}_{-k}$ of the form

$$f = (\mathscr{D}^*)^k \mathscr{D}^k \Phi \text{ for some } \Phi \in \mathscr{S}_{k,0} ,$$

satisfies

$$(f, \Psi) = (R_k f, \Psi)_{\mathscr{S},k} = (\Phi,\Psi) = 0 \text{ for all } \Psi \in \mathscr{S}^{\perp}_{k,0}.$$

□

Let now $S_{k,0}$ be the $\mathscr{S}_k$-orthonormal projection onto $\mathscr{S}_{k,0}$ and $S_{-k,0}$ the $\mathscr{S}_{-k}$-orthogonal projection on $\mathscr{S}_{-k,0}$. Furthermore let

$$T_{k,0} = (1 - S_{k,0}) , k \in \mathbb{Z}. \tag{1.184}$$

Then, collecting our results we have

Theorem 1.8: *For the spaces* $\mathscr{S}_k$, $\mathscr{S}_{-k}$, $k \in \mathbb{N} \cup \{0\}$, *the following orthogonal decompositions are valid*

$$\mathscr{S}_k = \mathscr{S}_{k,0} \oplus \mathscr{S}^{\perp}_{k,0} ,$$

$$\mathscr{S}_{-k} = \mathscr{S}_{-k,0} \oplus \mathscr{S}^{\perp}_{-k,0} .$$

Elements in $\mathscr{S}^{\perp}_{k,0}$ *can be approximated by functions in* $C_\infty(\mathbb{R}-\{0\})$.
Moreover, we have

$$\mathscr{S}_{k,0} = R_k \mathscr{S}_{-k,0}, \ \mathscr{S}^{\perp}_{k,0} = R_k \mathscr{S}^{\perp}_{-k,0}$$

and

$$\mathscr{S}_{-k,0} = \text{span } \{D^s \delta \mid s \in \mathbb{N} \cup \{0\}, s < k \}.$$

Between the corresponding orthogonal projections we have the relationship

$$S_{-k,0} = (\mathscr{D}^*)^k \mathscr{D}^k S_{k,0} R_k , \ T_{-k,0} = (\mathscr{D}^*)^k \mathscr{D}^k T_{k,0} R_k.$$

The projections $S_{-k,0}$, $T_{-k,0}$ *are dual to* $S_{k,0}$, $T_{k,0}$ *in the sense that*

$$(S_{-k,0} f, \Phi) = (f, S_{k,0} \Phi) , \ (T_{-k,0} f, \Phi) = (f, T_{k,0} \Phi) ,$$

for all $f \in \mathscr{S}_{-k}$ *and* $\Phi \in \mathscr{S}_k$, $k \in \mathbb{N} \cup \{0\}$.

Proof: Except maybe for the statements about projections, the results are clear from the above considerations for $k \in \mathbb{N}$ and the case $k = 0$ is trivial. The results about the projections are also easy to obtain. Let $f \in \mathscr{S}_{-k,0}$, $k \in \mathbb{N}$, then

$$T_{-k,0}\, f = 0,$$

and equivalently

$$(f,\ T_{k,0}\, \Phi) = 0.$$

From this the duality statement about $S_{k,0}$ and $S_{-k,0}$ is obvious. Moreover,

$$\begin{aligned}(S_{-k,0}\, f,\ \Phi) &= (f,\ S_{k,0}\, \Phi) = (R_k\, f,\ S_{k,0}\, \Phi)_{\mathcal{S},k'} \\ &= (S_{k,0}\, R_k\, f,\ \Phi)_{S,k} = (R_k\, (\mathcal{D}^*)^k\, \mathcal{D}^k\, S_{k,0}\, R_k\, f, \Phi)_{\mathcal{S},k'} \\ &= ((\mathcal{D}^*)^k\, \mathcal{D}^k\, S_{k,0}\, R_k\, f, \Phi).\end{aligned}$$

The remaining statements about $T_{-k,0}$ and $T_{k,0}$ can be obtained similarly or by using the definition (1.184).

□

As an example, we have that the Dirac δ distribution is the only linear independent element of $\mathcal{S}_{-1,0}$.

Because of the continuous imbedding of $\mathcal{S}_{k+1}$ in $\mathcal{S}_k$ it is clear that $\mathcal{S}^{\perp}_{k+1,0} \subset \mathcal{S}^{\perp}_{k,0}$; also, $\mathcal{S}_{-k,0} \subset \mathcal{S}_{-k-1,0}$, $k \in \mathbb{N} \cup \{0\}$. This implies a certain independence of $k \in \mathbb{N} \cup \{0\}$.

Using the projections $T_{k,0}$ associated with $\mathcal{S}_{k,0}$ we may define a modified 'multiplier' $Sgn_k(\boldsymbol{m})$ based on the sgn function:

$$Sgn_k(\boldsymbol{m}) = sgn(\boldsymbol{m})\ T_{k,0} \text{ on } \mathcal{S}_k.$$

Then, in keeping with the definition strategy displayed in (1.107) - (1.109) we define :

$$(Sgn_{-k}(\boldsymbol{m})\ f, \Phi) := (f\ ,\ Sgn_k(\boldsymbol{m})\ \Phi) \tag{1.185}$$

for $f \in \mathcal{S}_{-k}$ and $\Phi \in \mathcal{S}_k$, $k \in \mathbb{N} \cup \{0\}$.

With regard to (1.115) we have the important difference between $Sgn_k(\boldsymbol{m})$ and $sgn(\boldsymbol{m})$ that e.g.

$$D(sgn(\boldsymbol{m})\ \Phi) = sgn(\boldsymbol{m})\ D\Phi + \Phi(0),\ \Phi \in \mathcal{S}_1,$$

but

$$D(Sgn_1(\boldsymbol{m})\ \Phi) = Sgn_1(\boldsymbol{m})\ D\Phi\ ,\ \Phi \in \mathcal{S}_1.$$

By introducing the projection $T_{k,0}$ we have regularized $sgn(\boldsymbol{m})$ in the sense that the effect of the jump which results in an additional δ functional appearing in the derivative is removed. In fact we have — as a simple calculation shows — that

$$\mathrm{Sgn}_k(\mathscr{m})\, \mathscr{D}^s = \mathscr{D}^s\, \mathrm{Sgn}_k(\mathscr{m}),\ \mathrm{Sgn}_k(\mathscr{m})\, D^s = D^s\, \mathrm{Sgn}_k(\mathscr{m}) , \tag{1.186}$$

on $\mathscr{S}_k$, $k,s \in \mathbb{N}$, $s \le k$.

Both multipliers $\mathrm{Sgn}_k(\mathscr{m})$ and $\mathrm{sgn}(\mathscr{m})$ are unitary mappings on $\mathscr{S}^{\perp}_{k,0}$, $k \in \mathbb{N} \cup \{0\}$, in fact they are equal on this subspace. For the extended multiplier $\mathrm{Sgn}_{-k}(\mathscr{m})$, we immediately have from the definition (1.185)

$$T_{-k,0}\, \mathrm{Sgn}_{-k}(\mathscr{m}) = \mathrm{Sgn}_{-k}(\mathscr{m}) = \mathrm{Sgn}_{-k}(\mathscr{m})\, T_{-k,0} \text{ on } \mathscr{S}_{-k} \tag{1.187}$$

in particular,

$$\mathrm{Sgn}_{-k}(\mathscr{m})\, f = 0 \text{ for } f \in \mathscr{S}_{-k,0} ,\ k \in \mathbb{N}. \tag{1.188}$$

On the other hand if $f \in \mathscr{S}_{-k}$ is such that

$$\mathrm{Sgn}_{-k}(\mathscr{m})\, f = 0,$$

then it follows that

$$\mathrm{supp}(f) = \{0\}.$$

In other words, the null space of $\mathrm{Sgn}_{-k}(\mathscr{m})$ as a mapping in $\mathscr{S}_{-k}$ is $\mathscr{S}_{-k,0}$:

$$\mathscr{S}_{-k,0} = \{f \in \mathscr{S}_{-k} \mid \mathrm{Sgn}_{-k}(\mathscr{m})\, f = 0\}. \tag{1.189}$$

In particular,

$$(\mathrm{Sgn}_{-k}(\mathscr{m}))^2 = T_{-k,0}.$$

There is another more common possibility to extend $\mathrm{sgn}(\mathscr{m})$,(to rather special functionals in $\mathscr{S}_{-k}$, however). The multiplier sgn(m) can be defined for any $f \in \mathscr{S}_{-k}$, $k \in \mathbb{N}$, with $0 \notin \mathrm{supp}(f)$. Since $0 \notin \mathrm{supp}(f)$ there is a neighbourhood of zero which does not belong to supp(f). Let $\alpha \in C_\infty(\mathbb{R})$ have support in this neighbourhood and be such that $\alpha \equiv 1$ near zero. Then we define

$$(\mathrm{sgn}(\mathscr{m})\, f,\ \Phi) := (f,\ \mathrm{sgn}(\mathscr{m})\ (1-\alpha)\Phi) \tag{1.190}$$

for $\Phi \in \mathscr{S}_k$. Note that $(1-\alpha)\ \Phi \in \mathscr{S}^{\perp}_{k,0}$ for $\Phi \in \mathscr{S}_k$ and that this definition does not depend on the particular choice of $\alpha \in C_\infty(\mathbb{R})$ as long as it satisfies the above requirements. Definition (1.190) describes the usual extension of the $\mathrm{sgn}(\mathscr{m})$ multiplier to generalized functions.

To translate our findings into properties of the Hilbert transform, we first define the extension of sgn(D) with the understanding of the extension

$Sgn_{-k}(\mathbf{m})$ as developed so far, i.e.

$$Sgn_{-k}(D) = \mathcal{F}^* Sgn_{-k}(\mathbf{m})\ \mathcal{F} \text{ on } \mathcal{S}_{-k}\ ,\ k \in \mathbb{N}.$$

From the corresponding property of $Sgn_{-k}(\mathbf{m})$ we have

$$Sgn_{-k}(D) \text{ is a unitary operator on } (\mathcal{D}^*)^k\ \mathcal{D}^k\ \mathcal{F}^*\ \mathcal{S}^{\perp}_{k,0},$$

as a subspace of $\mathcal{S}_{-k}$, $k \in \mathbb{N}$.

A more accessible characterization of $(\mathcal{D}^*)^k\ \mathcal{D}^k\ \mathcal{F}^*\ \mathcal{S}^{\perp}_{k,0}$ is obtained from the orthogonal decomposition (1.184):

$$Sgn_{-k}(D) \text{ is unitary}$$

$$\text{on } \{\Phi \mid \Phi \text{ is polynomial of degree } d < k\}^{\perp},$$

here the ortho-complement is of course taken in the sense of $\mathcal{S}_{-k}$, $k \in \mathbb{N}$. Note in particular that those polynomials are in fact elements of $\mathcal{S}_{-k}$ as is implicit in the above considerations. Using (1.190) we can also extend

$$sgn(D) \text{ defined on } \mathcal{F}^*\{f \in \mathcal{S}_{-k} \mid 0 \notin supp(f)\}.$$

The domain of this operator is not as easy to describe. However, the extension $Sgn_{-k}(\mathbf{m})$ has its own particular awkwardness in the use of the projection $T_{k,0}$. Since $T_{0,0}$ is the identity, we can combine the two ideas of extension in a simple way

$$sgn(\mathbf{m})\ f = Sgn_0(\mathbf{m})\ \alpha\ f\ + sgn(\mathbf{m})\ (1-\alpha)\ f, \tag{1.191}$$

for $f \in \mathcal{S}_{-k}$ such that $\alpha\ f \in \mathcal{S}_0 \subset \mathcal{S}_{-k}$. Here $\alpha \in C_{\infty}(\mathbb{R})$ is such that $\alpha \equiv 1$ near zero. Again this definition is independent of α provided $f \in \mathcal{S}_{-k}$ is such that $\alpha\ f \in \mathcal{S}_0$.

Recalling that the Heaviside function $h = \frac{1}{2}(1 + sgn(\mathbf{m}))$ and that the characteristic function $\Pi = (\sigma_{-1/2} - \sigma_{1/2})\ sgn(\mathbf{m})$, we learn that h and Π can be extended in similar ways, either in the sense of (1.190), (1.191) —— in which case we keep the same notation —— or in the sense of (1.185). In the latter case, we shall use a different notation in order to avoid confusion :

$$H_{-k} = \frac{1}{2}(1 + Sgn_{-k}(\mathbf{m})) \text{ and } \Pi_{-k} = (\sigma_{-1/2} - \sigma_{1/2})\ Sgn_{-k}(\mathbf{m}). \tag{1.192}$$

As an application we note the following obvious consequence of the setting developed above:

$$\text{For } f \in \mathscr{S}_k : H_k f = f \text{ iff } \mathrm{Sgn}_k(D)\, \mathscr{F}^* f = \mathscr{F}^* f,\ k \in \mathbb{Z}.$$

A similar result holds for sgn(D), h extended in the sense of (1.190). For $f \in \mathscr{S}_k$ with $0 \notin \mathrm{supp}(f)$, $k \in \mathbb{Z}$, we have:

$$hf = f \text{ iff } \mathrm{sgn}(D)\, \mathscr{F}^* f = \mathscr{F}^* f.$$

In view to (1.191) we can also formulate: For $f \in \mathscr{S}_k$, $k \in \mathbb{Z}$, such that $\alpha f \in \mathscr{S}_0$ for some $\alpha \in C_\infty(\mathbb{R})$ with $\alpha \equiv 1$ in a neighbourhood of zero we have

$$hf = f \text{ iff } \mathrm{sgn}(D)\, \mathscr{F}^* f = \mathscr{F}^* f. \tag{1.193}$$

Function with the property on the left-hand side of (1.193) are called 'causal', so (1.193) states that eigensolutions of the Hilbert transform for eigenvalue 1 are all those functions with a causal Fourier transform (compare e.g. [26], p. 198 pp).

<u>1.1.3.2.2 Abel transform.</u>

Let us consider the complex-valued multiplier $\mathbf{m}^\alpha$ defined by

$$(\mathbf{m}^\alpha f)(t) = e^{\alpha\pi i(1-\mathrm{sgn}(t))/2}\, |t|^\alpha f(t),\ \alpha \in \mathbb{R}, \tag{1.194}$$

for $f \in L_2(\mathbb{R})$ such that $|\mathbf{m}|^\alpha f \in L_2(\mathbb{R})$. Since for $\alpha \in \mathbb{Z}$ we have

$$(\mathbf{m}^\alpha f)(t) = e^{\alpha\pi i(1-\mathrm{sgn}(t))/2}\, |t|^\alpha f(t) = (\mathrm{sgn}(t))^\alpha\, |t|^\alpha f(t) = t^\alpha f(t),$$

definition (1.194) generalizes powers with integer exponent. The spectral theorem implies for the corresponding operators D^α that they are normal as (densely defined) operators on $L_2(\mathbb{R})$ We have of course

$$D(D^\alpha) = \{\, \Phi \in L_2(\mathbb{R}) \mid |\mathbf{m}|^\alpha \mathscr{F}\Phi \in L_2(\mathbb{R}) \,\}.$$

For $\alpha \in (-1,0)$ operator D^α is called fractional integral. The Abel transform is closely related to the special fractional integral for $\alpha = -\frac{1}{2}$. From the definition we have that

$$D^{-1/2}\, D^{-1/2}\, D = \mathrm{id} \text{ on } H_1(\mathbb{R}),$$

which shows that $D^{-1/2}$ is a sort of 'half integral'. In order to write $D^{-1/2}$ as a convolution we have to calculate the Fourier transform of $\mathbf{m}^{-1/2}$. First we note that $\mathfrak{m}^{-1/2}$ is in $\mathcal{S}_{-1}$. For any $\Phi \in C_\infty(\mathbb{R})$ we have

$$(\mathfrak{m}^{-1/2}, \Phi) = \int_{\mathbb{R}} t^{-1/2} \Phi(t)\, dt.$$

Thus,

$$|(\mathfrak{m}^{-1/2}, \Phi)| \le \int_{\mathbb{R}} |t|^{-1/2} |\Phi(t)|\, dt,$$

$$\le \left\{ \int_{|t|\ge 1} (|t|\,(1+|t|^2))^{-1}\, dt \right\}^{1/2} \left\{ \int_{\mathbb{R}} (1+|t|^2)\, |\Phi(t)|^2\, dt \right\}^{1/2} +$$

$$+ \int_{|t|\le 1} |t|^{-1/2}\, dt \sup_{|x|\le 1} |\Phi(x)|,$$

$$\le C\, \|\Phi\|_{\mathcal{S},1}.$$

For the last step we again have applied Sobolev estimate (1.113),

$$|\Phi(x)| = |(\sigma_{-x}\Phi)(0)| \le 2^{-1/4}\pi\, \|\mathfrak{D}\,\sigma_{-x}\Phi\|,$$

$$\le 2^{-1/4}\pi\, \|\sigma_{-x}(\mathfrak{D}\,\Phi) - i\,x\,\sigma_{-x}\Phi\|,$$

$$\le 2^{-1/4}\pi\, \|\sigma_{-x}(\mathfrak{D}\,\Phi)\| + |x|\, \|\sigma_{-x}\Phi\|,$$

$$\le 2^{-1/4}\pi\, \|\mathfrak{D}\,\Phi\| + |x|\, \|\Phi\|,$$

$$\le \max(2^{-1/4}\pi, |x|)\, \|\Phi\|_{\mathcal{S},1}.$$

Thus, we see indeed that $\mathfrak{m}^{-1/2} \in \mathcal{S}_{-1}$. Estimate (1.195) also shows that $|\mathfrak{m}|^{-1/2}$ is in $\mathcal{S}_{-1}$. Using the characteristic function Π as a cut-off function we note that

$$(e_{\alpha,0}\,\Pi)\, |\mathfrak{m}|^{-1/2},\; (e_{\alpha,0}\,\Pi)\, \mathfrak{m}^{-1/2} \in L_1(\mathbb{R}),\; \alpha \ne 0.$$

An analogous estimate to (1.195) not only shows that

$$(e_{\alpha,0}\,\Pi)\, |\mathfrak{m}|^{-1/2},\; (e_{\alpha,0}\,\Pi)\, \mathfrak{m}^{-1/2} \in \mathcal{S}_{-1},\; \alpha \ne 0,$$

uniformly in α, but also that

$$(\ell_{\alpha,0}\,\Pi)\,|m|^{-1/2} \longrightarrow |m|^{-1/2} \text{ in } \mathcal{S}_{-1} \text{ as } \alpha \longrightarrow 0+\,,$$

$$(\ell_{\alpha,0}\,\Pi)\,m^{-1/2} \longrightarrow m^{-1/2} \text{ in } \mathcal{S}_{-1} \text{ as } \alpha \longrightarrow 0+. \tag{1.196}$$

With Remark 1.11 (ii) in mind we calculate

$$\begin{aligned}
\mathcal{F}((\ell_{\alpha,0}\,\Pi)\,|m|^{-1/2})(t) &= \int_{-1/\alpha}^{1/\alpha} e^{-2\pi i t y}\,|y|^{-1/2}\,dy\,, \\
&= \int_{1/\alpha}^{1/\alpha} \cos(2\pi\, ty)\,|y|^{-1/2}\,dy\,, \\
&= |t|^{-1/2} \int_{-|t|/\alpha}^{|t|/\alpha} \cos(2\pi\, u)\,|u|^{-1/2}\,du\,,
\end{aligned}$$

for $\alpha > 0$. Thus, letting $\alpha \longrightarrow 0+$ we obtain with (1.196) that

$$\mathcal{F}\,|m|^{-1/2} = |m|^{-1/2} \int_{\mathbb{R}} \cos(2\pi\, u)\,|u|^{-1/2}\,du. \tag{1.197}$$

That the latter integral exists as an improper integral follows from the fact, that due to the periodicity of the trigonometric functions the integral can be written as an alternating series of vanishing terms. Following an idea in [34], p. 181, we apply the same reasoning to the right hand side of (1.171) that was employed to obtain (1.197). Thus,

$$\begin{aligned}
|m|^{-1/2} &= \mathcal{F}\,\mathcal{F}\,|m|^{-1/2} = \mathcal{F}\,(|m|^{-1/2} \int_{\mathbb{R}} \cos(2\pi\, u)\,|u|^{-1/2}\,du) \\
&= |m|^{-1/2} \left\{ \int_{\mathbb{R}} \cos(2\pi\, u)\,|u|^{-1/2}\,du \right\}^2.
\end{aligned}$$

This determines the value of the improper integral

$$\int_{\mathbb{R}} \cos(2\pi\, u)\,|u|^{-1/2}\,du = 1. \tag{1.198}$$

So, (1.198) and (1.197) together yield

$$\mathcal{F}\,|m|^{-1/2} = |m|^{-1/2}. \tag{1.199}$$

An analogous investigation of $|m|^{-1/2}\,\mathrm{sgn}(m)$ leads to

$$\mathcal{F}\,(|m|^{-1/2}\,\mathrm{sgn}(m)) = -i\,|m|^{-1/2}\mathrm{sgn}(m), \tag{1.200}$$

and

$$\int_{\mathbb{R}} \sin(2\pi\,|u|)\,|u|^{-1/2}\,du = 1. \tag{1.201}$$

By an analogous reasoning we can calculate the Fourier transform of $m^{-1/2}$.

$$\mathcal{F}((\ell_{\alpha,0}\,\Pi)\,m^{-1/2})(t) = \int_{-1/\alpha}^{1/\alpha} e^{-2\pi ity}\,y^{-1/2}\,dy\,,$$

$$= \int_{0}^{1/\alpha} e^{-2\pi ity}\,y^{-1/2}\,dy + i\int_{-1/\alpha}^{0} e^{-2\pi ity}\,|y|^{-1/2}\,dy,$$

$$= |t|^{-1/2}\left\{\int_{0}^{|t|/\alpha} e^{-2\pi i\,\mathrm{sgn}(t)s}\,s^{-1/2}\,ds + i\int_{0}^{|t|/\alpha} e^{2\pi i\,\mathrm{sgn}(t)s}\,s^{-1/2}\,ds\right\}$$

$$= (1+i)\,|t|^{-1/2}\left\{\int_{0}^{|t|/\alpha} (\cos(2\pi\,s) - \mathrm{sgn}(t)\,\sin(2\pi s))\,s^{-1/2}\,ds\right\}.$$

By letting $\alpha \longrightarrow 0+$ and making use of (1.198) and (1.201) we get

$$\mathcal{F}\,m^{-1/2} = (1+i)\,|m|^{-1/2}\,(1-h) = (1+i)\,|m|^{-1/2}\,\frac{1}{2}\,(1 + \mathrm{sgn}(m)).$$

Thus,

$$\mathcal{F}^{*}\,m^{-1/2} = (1+i)\,|m|^{-1/2}\,h = (1+i)\,|m|^{-1/2}\,\frac{1}{2}\,(1 - \mathrm{sgn}(m)) \tag{1.202}$$

and so

$$D^{-1/2}\,f = (1+i)\,(|m|^{-1/2}\,h) * f\,, \tag{1.203}$$

$$(D^{-1/2})^{*}\,f = (1-i)\,(|m|^{-1/2}\,(1-h)) * f\,, \tag{1.204}$$

for all $f \in \mathcal{S}_1$.

For —— say —— $f \in C_\infty(\mathbb{R})$ we can write the real part of $(D^{-1/2})^{*}$ as an ordinary integral

$$((|\boldsymbol{m}|^{-1/2}(1-h)) * f)(t^2) = 2\int_t^\infty \frac{f(r^2)\, r}{(r^2 - t^2)^{1/2}}\, dr\ ,\ t \geq 0.$$

This interpreted as a transformation of $g(x) \equiv f(x^2)$ is known under the name Abel transform.

The real convolution operators $K_\pm$ defined by

$$K_+ f = \pi^{-1/2}(|\boldsymbol{m}|^{-1/2} h)) * f = \frac{1-i}{2}\pi^{-1/2} D^{-1/2} f,$$

$$K_- f = \pi^{-1/2}(|\boldsymbol{m}|^{-1/2}(1-h)) * f = \frac{1+i}{2}\pi^{-1/2}(D^{-1/2})^* f,$$

for f in the range of $|D|^{1/2}$ (as a selfadjoint operator on $L_2(\mathbb{R})$), have the noteworthy properties

$$K_+^2 f' = D^{-1/2} D^{-1/2} D f = f,$$

$$K_-^2 f' = -(D^{-1/2})^* (D^{-1/2})^* D f = -f,$$

for $f \in H_1(\mathbb{R})$.

We shall now briefly discuss the possibility of extending the meaning of $D^{-1/2}$ to spaces $\mathscr{S}_{-k}$, $k \in \mathbb{N}$. In order to extend $D^{-1/2}$ we first notice that by definition

$$\boldsymbol{m}^{-1/2} = |\boldsymbol{m}|^{-1/2}(h + i(1-h)) = \frac{1+i}{2}|\boldsymbol{m}|^{-1/2}(1 - i\,\mathrm{sgn}(\boldsymbol{m})).$$

Since the point zero is again problematic, it seems to be suitable to use the ideas of the previous section. Since extension in the sense of (1.191) is fairly obvious we consider the extension with the projection technique and define

$$(\boldsymbol{m}^{-1/2})_k = \frac{1+i}{2}|\boldsymbol{m}|^{-1/2}(1 - i\,\mathrm{sgn}(\boldsymbol{m}))\, T_{k,0}\ ,\ k \in \mathbb{N}.$$

That this is meaningful has to be seen yet. We shall demonstrate that in fact

$$(\boldsymbol{m}^{-1/2})_k : \mathscr{S}_k \longrightarrow \mathscr{S}_{k-1}, \qquad (1.205)$$

is a bounded linear mapping , $k \in \mathbb{N}$. The observation that

$$(1 - i\,\mathrm{sgn}(\boldsymbol{m}))\, T_{k,0} : \mathscr{S}_k \longrightarrow \mathscr{S}_k\ ,\ k \in \mathbb{N} \cup \{0\},$$

is bounded and maps into $\mathscr{S}_{k,0}^{\perp}$ reduces the problem to considering $|\boldsymbol{m}|^{-1/2}$ $T_{k,0}$. So, let $\Phi \in \mathscr{S}_{k,0}^{\perp}$. By the density of $C_\infty(\mathbb{R}-\{0\})$ in this subspace and by using Sobolev's inequality (Lemma 1.2) we see that

$$(D^s \Phi) \restriction_0 = 0 \text{ for } s \in \mathbb{N} \cup \{0\},\ s < k.$$

This yields —— via Taylor's expansion and again Sobolev's lemma —— the estimate

$$\left| \, |x|^{j-s-1/2} \, (D^{k-s-j-1} \Phi)(x) \, \right| \leq C \, |x|^{k-s-1} \, \|\Phi\|_{\mathcal{S},k} \, ,$$

for some generic constant C independent of $\Phi \in \mathcal{S}^{\perp}_{k,0}$, $s \in \mathbb{N} \cup \{0\}$, $s < k$, in a neighbourhood of zero —— say $|x| \leq 1$. With this point-wise estimate in a neighbourhood of zero and since $|x|^{-1} \leq 1$ for $|x| \geq 1$ we have

$$\| \, |\boldsymbol{m}|^{-1/2} \Phi \|_{\mathcal{S},k-1} \leq C \, \|\Phi\|_{\mathcal{S},k} \text{ for all } \Phi \in \mathcal{S}^{\perp}_{k,0},\ k \in \mathbb{N}.$$

This estimate finally proves (1.205).

Thus, similar as before we can define an extension $(\boldsymbol{m}^{-1/2})_{-k+1}$ via

$$((\boldsymbol{m}^{-1/2})_{-k+1} \, f, \, \Phi) := (f, \, (\boldsymbol{m}^{-1/2})_k \, \Phi)$$

for all $f \in \mathcal{S}_{-k+1}$, $\Phi \in \mathcal{S}_k$, $k \in \mathbb{N}$, and

$$(D^{-1/2})_{-k+1} := \mathcal{F}^* \, (\boldsymbol{m}^{-1/2})_{-k+1} \, \mathcal{F} : \mathcal{S}_{-k+1} \longrightarrow \mathcal{S}_{-k}. \tag{1.206}$$

<u>1.1.3.3 The sampling theorem for the Fourier transform.</u>

Let us reconsider the functional

$$Ш = \sum_{n \in \mathbb{Z}} \sigma_n \, \delta \in \mathcal{S}_{-3}, \tag{1.207}$$

(see (1.118)). We want to calculate its Fourier transform. As a first step in this direction let us consider the finite sums

$$Ш_N = \sum_{n=-N}^{N} \sigma_n \, \delta \in \mathcal{S}_{-1},\ n \in \mathbb{N}. \tag{1.208}$$

We have $Ш_N \longrightarrow Ш$ in $\mathcal{S}_{-3}$ as $N \longrightarrow \infty$. From Theorem 1.4 we have

$$\mathcal{F}^* \, Ш_N = \sum_{n=-N}^{N} e^{2\pi i n \boldsymbol{m}} .$$

The operators

$$Ш_N(D) = \mathcal{F}^* \, Ш_N \, \mathcal{F} \, ,\ N \in \mathbb{N},$$

are well-defined as bounded linear operators on $\mathcal{S}_k$ for any $k \in \mathbb{Z}$. Using the convolution notation, we get

$$Ш_N(D)\ f = \Delta_N * f \text{ for } f \in \mathcal{S}_k,\ k \in \mathbb{Z},\ N \in \mathbb{N}, \tag{1.208}$$

with

$$\Delta_N = \sum_{n=-N}^{N} e^{2\pi i n \boldsymbol{m}} = 1 + 2 \sum_{n=0}^{N} \cos(2\pi n \boldsymbol{m}). \tag{1.209}$$

Note that so far the functions of D we considered have all been Borel functions. But here we discuss a distribution of D. Using that

$$2 \sin(\pi\ x)\ \cos(2\pi nx) = \sin(2\pi(n + \tfrac{1}{2})x) - \sin(2\pi(n - \tfrac{1}{2})x),$$

we have

$$2 \sin(\pi\ x) \sum_{n=0}^{N} \cos(2\pi nx) = \sum_{n=0}^{N} (\sin(2\pi(n + \tfrac{1}{2})x) - \sin(2\pi(n - \tfrac{1}{2})x)),$$

$$= \sin(2\pi(N+\tfrac{1}{2})x) - \sin(\pi x).$$

Thus,

$$\Delta_N = \frac{\sin(\pi(2N+1)x)}{\sin(\pi x)},\quad x \notin \mathbb{Z}. \tag{1.209}$$

For $x \in \mathbb{Z}$ we can take the continuous extension of (1.209) to have a complete description of Δ_N, $N \in \mathbb{N}$. The functions Δ_N, $N \in \mathbb{N}$, are known as Dirichlet kernels. For e.g. $f \in \mathcal{S}_1$ we can easily calculate

$$\Delta_N * f = \mathcal{F}^* Ш_N \mathcal{F} f = \mathcal{F}^* \sum_{n=-N}^{N} (e^{2\pi i n \boldsymbol{m}}, f)\ \sigma_n\ \delta = \sum_{n=-N}^{N} (e^{2\pi i n \boldsymbol{m}}, f)\ e^{2\pi i n \boldsymbol{m}}.$$

If f has support in $(-1/2,1/2)$, $\Delta_N * f$ apparently features an approximation of the Fourier series of f and thus a trigonometric approximation of f itself. Let $\varphi \in C_\infty(\mathbb{R})$ be such that $\mathrm{supp}(\varphi) \subset (-1/2,1/2)$ and $\varphi \equiv 1$ in a neighbourhood of zero. Then define the function

$$\alpha := \sum_{n\in\mathbb{Z}} \sigma_n\ \varphi \in \mathcal{S}_{-1}.$$

Since α and all its derivatives are bounded, it maps — used as a multiplier — $\mathcal{S}_k$ into itself, $k \in \mathbb{N} \cup \{0\}$. Now consider a $\Phi \in \mathcal{S}_3$ then $\alpha\ \Phi \in \mathcal{S}_3$ and $(1-\alpha)\ \Phi \in \mathcal{S}_3$ and of course $\Phi = \alpha\ \Phi + (1-\alpha)\ \Phi$. This induces a decomposition of $f \in \mathcal{S}_{-3}$: $f = \alpha\ f + (1-\alpha)\ f$. We shall deal with the corresponding two parts of Δ_N separately.

$$(\alpha\, \Delta_N, \Phi) \equiv (\Delta_N, \alpha\, \Phi) = \sum_{n=-N}^{N} (e^{2\pi i n \mathbf{m}}, \alpha\, \Phi). \tag{1.210}$$

Making use of

$$e^{2\pi i n \mathbf{m}} = \frac{1}{n} D\, e^{2\pi i n \mathbf{m}}, n \in \mathbb{N},\ n \neq 0,$$

we obtain

$$(e^{2\pi i n \mathbf{m}}, \alpha\, \Phi) = n^{-2}\, (e^{2\pi i n \mathbf{m}}, D^2(\alpha\, \Phi)).$$

Thus we get an estimate of the kind

$$|(e^{2\pi i n \mathbf{m}}, \alpha\, \Phi)| \leq C\, n^{-2}\, \|\Phi\|_{\mathscr{S},3}. \tag{1.211}$$

Thus, $\{\alpha\, \Delta_N\}_N$ is in fact a Cauchy sequence in $\mathscr{S}_{-3}$. Indeed, let $N \leq M$ then

$$|(\alpha\, \Delta_N - \alpha\, \Delta_M, \Phi)| \equiv |(\Delta_N - \Delta_M, \alpha\, \Phi)| \leq C\, (\sum_{n=N}^{M} n^{-2})\, \|\Phi\|_{\mathscr{S},3}. \tag{1.212}$$

The other part of Δ_N can be estimated in a similar way using (1.209).We have

$$|((1-\alpha)\, \Delta_N, \Phi)| = |(\frac{\sin(\pi(2N+1)\mathbf{m})}{\sin(\pi \mathbf{m})}, (1-\alpha)\, \Phi)|,$$

$$\leq (N+\tfrac{1}{2})^{-2}\, |(\sin(\pi(2N+1)\mathbf{m}, D^2\, (\frac{(1-\alpha)}{\sin(\pi\mathbf{m})}\, \Phi))| \tag{1.213}$$

$$\leq C\, (N+\tfrac{1}{2})^{-2}\, \|\Phi\|_{\mathscr{S},3}.$$

In the last step we have made use of the periodicity of $\sin(\pi\mathbf{m})$ and the uniform boundedness of $\frac{1}{\sin(\pi\mathbf{m})}$ on $\text{supp}(1-\alpha)$ (Note: $0 \notin \text{supp}(1-\varphi)$). Taking this estimate and (1.212) we obtain

$$|(\Delta_N - \Delta_M, \Phi)| \leq C \left\{ (\sum_{n=N}^{M} n^{-2}) + (N+\tfrac{1}{2})^{-2} + (M+\tfrac{1}{2})^{-2} \right\} \|\Phi\|_{\mathscr{S},3} \tag{1.214}$$

Since $\{\Delta_N\}_N$ is a Cauchy sequence in $\mathscr{S}_{-3}$, it converges to an element $T \in \mathscr{S}_{-3}$. According to (1.213) $|(\Delta_N, \Phi) - (\Delta_N, \alpha\, \Phi)|$ tends to zero, thus it suffices to consider $(\Delta_N, \alpha\, \Phi)$. Let $\Phi \in \overset{\circ}{C}_\infty(\mathbb{R})$ then

$$(\Delta_N, \alpha\, \Phi) = \sum_{n\in\mathbb{Z}} (\Delta_N, (\sigma_n\, \alpha)\, \Phi) = \sum_{n\in\mathbb{Z}} (\Delta_N, \sigma_n\, (\alpha\, \sigma_{-n}\, \Phi)),$$

$$= \sum_{n\in\mathbb{Z}} (\Delta_N, \alpha\, \sigma_{-n}\, \Phi).$$

Note that the sum is in fact finite due to the assumption that $\Phi \in C_\infty(\mathbb{R})$. Each term $(\Delta_N, \alpha\, \sigma_{-n} \Phi)$ is the partial sum of the Fourier series of $\sigma_{-n} \Phi$ evaluated at zero, thus it follows that $T = Ш$ and we have shown that

$$\mathcal{F}\, Ш = Ш \, . \tag{1.215}$$

Remark 1.14: *We also have shown that*

$$\Delta_N \longrightarrow Ш \quad \textit{in } \mathcal{S}_{-3} \textit{ as } N \longrightarrow \infty \, .$$

Of course, Δ_N *is not the only possible choice to approximate* $Ш$*. Any other periodic extension of a* δ*-family could be used. Among the more well-known approximations of* $Ш$ *are the so-called Fejer kernels*

$$F_N = (N+1)^{-1} \sum_{n=0}^{N} \Delta_n = \frac{1-\cos(2\pi(N+1)\mathbf{m})}{2\,(\sin(\pi\mathbf{m}))^2} = \left\{\frac{\sin(\pi(N+1)\mathbf{m})}{\sin(\pi x)}\right\}^2, \quad N \in \mathbb{N},$$

(see e.g. [38]*, p. 212 pp).*

Based on (1.215) we can now discuss $Ш(D)$. The definition of $Ш(D)$ is given in the usual way

$$Ш(D) = \mathcal{F}^{*}\, Ш\, \mathcal{F} \tag{1.216}$$

and its domain can be chosen e.g. as the space $\mathcal{S}_1$

$$Ш(D) : \mathcal{S}_1 \longrightarrow \mathcal{S}_{-3}.$$

This becomes transparent if we note that

$$Ш\, f = Ш\, \Big(\sum_{n\in\mathbb{Z}} f(n)\, \sigma_n \varphi\Big), \text{ for } f \in \mathcal{S}_1 \, ,$$

where φ is as above. The function $\big(\sum_{n\in\mathbb{Z}} f(n)\, \sigma_n \varphi\big) \in \mathcal{S}_{-1}$ is an interpolant of f and is with all its derivatives bounded. Thus, similarly as before

$$\Big(\sum_{n\in\mathbb{Z}} f(n)\, \sigma_n \varphi\Big)\, \Phi \in \mathcal{S}_3 \quad \text{for } \Phi \in \mathcal{S}_3 \, .$$

Noting that

$$(Ш\, f, \Phi) = \Big(\Big(\sum_{n\in\mathbb{Z}} f(n)\, \sigma_n \varphi\Big)\, Ш, \Phi\Big) = \Big(Ш, \Big(\sum_{n\in\mathbb{Z}} f(n)\, \sigma_n \varphi\Big)\, \Phi\Big),$$

for $f \in \mathcal{S}_1$, $\Phi \in \mathcal{S}_3$, we see that $Ш(D)$ maps $\mathcal{S}_1$ into $\mathcal{S}_{-3}$.

In fact we see that we may choose $L_1(\mathbb{R})$ or any other set of functions that guarantees continuity and boundedness of the Fourier transform in a neighbourhood of $\mathbb{Z}$. If we weaken the mapping properties of $Ш$ we can extend even further. We shall not pursue this, however. Using the convolution notation we can write

$$Ш(D)\ \Phi \equiv Ш * \Phi \text{ for } \Phi \in \mathcal{S}_1.$$

To understand the action of $Ш(D)$ we consider the action of $Ш_N(D)$ on functions in $C_\infty(\mathbb{R})$.

$$\begin{aligned} Ш_N(D)\ \Phi &= \Delta_N * \Phi, \\ &= \sum_{n=-N}^{N} (e^{2\pi i n \mathbf{m}}, \Phi)\ e^{2\pi i n \mathbf{m}}. \end{aligned} \tag{1.217}$$

To investigate this further we use the partition of unity

$$1 = \sum_{n\in\mathbb{Z}} \sigma_n\ \Pi. \tag{1.218}$$

with this and making use of the periodicity of $e^{2\pi i n \mathbf{m}}$ we have

$$\begin{aligned} Ш_N(D)\ \Phi &= \sum_{m\in\mathbb{Z}} \sum_{n=-N}^{N} (e^{2\pi i n \mathbf{m}}, (\sigma_m\ \Pi)\ \Phi)\ e^{2\pi i n \mathbf{m}}, \\ &= \sum_{m\in\mathbb{Z}} \sum_{n=-N}^{N} (e^{2\pi i n \mathbf{m}}, \Pi\ \sigma_{-m}\ \Phi)\ e^{2\pi i n \mathbf{m}}. \end{aligned} \tag{1.219}$$

Note again that the sums are finite. Equation (1.219) shows that for $N \longrightarrow \infty$ $Ш_N(D)\ \Phi$ converges to the sum of several Fourier series expansions, thus

$$Ш_N(D)\ \Phi \longrightarrow Ш(D)\ \Phi = \sum_{m\in\mathbb{Z}} w_m \text{ as } N \longrightarrow \infty.$$

where w_m equals $\Pi\ \sigma_{-m}\ \Phi$ on $(-1/2,1/2)$ and is the periodic extension of $\sigma_{-m}\ \Phi$ from $(-1/2,1/2)$ to all of $\mathbb{R}$. In other words w_m is the periodic extension of Φ restricted to the period interval $(m - 1/2,\ m + 1/2)$. Thus, if $p_\#$ denotes the periodic extension of a function restricted to $(-1/2,1/2)$ then by the density of $C_\infty(\mathbb{R})$ in $\mathcal{S}_1$,

$$Ш(D) = \sum_{m\in\mathbb{Z}} p_\#\ \sigma_m \text{ on } \mathcal{S}_1. \tag{1.230}$$

In particular we have

<u>Theorem 1.9</u>: *For* $f \in \mathcal{S}_1$ *with* $\mathrm{supp}(f) \subset (-1/2,1/2)$ *we have*

$$\Pi\, Ш(D)\, f = f. \tag{1.231}$$

<u>Proof</u>: The result follows for $f \in \mathcal{S}_1$ from (1.230). If $\mathrm{supp}(f) \subset (-1/2,1/2)$ then $\sigma_m f = 0$ in the intervall $(-1/2,1/2)$ for all $m \in \mathbb{Z}$, $m \neq 0$. Thus by (1.230)

$$Ш(D)\, f = p_{\#}\, f.$$

Noting that $p_{\#} f$ is zero near integer points, we can apply the cut-off function Π and the first part follows. □

The theorem can of course be reformulated for $f = \mathcal{F}^* g$, $g \in \mathcal{S}_1$. Also, by simply applying the Fourier transform to (1.231) we obtain that

$$g = \mathcal{F}\, \Pi\, \mathcal{F}^*\, Ш\, g = \mathcal{F}^*\, \Pi\, \mathcal{F}\, Ш\, g = \Pi(D)\, Ш\, g, \tag{1.232}$$

which might be considered as an interpolation theorem. Indeed, evaluating $\Pi(D)\, Ш$ yields the second part of the next theorem.

<u>Theorem 1.10</u>: *Let* $g \in \mathcal{S}_1$ *be such that its Fourier transform has support in the interval* $(-1/2,1/2)$ *then*

$$\Pi(D)\, Ш\, g = g. \tag{1.233}$$

Moreover, we have

$$\Pi(D)\, Ш\, g = \sum_{n\in\mathbb{Z}} g(n)\, \sigma_n\, \mathrm{sinc}, \text{ for } g \in \mathcal{S}_1. \tag{1.234}$$

<u>Proof</u>: Noting that $\mathcal{F}\, \Pi\, \mathcal{F}^* = \mathcal{F}^*\, \Pi\, \mathcal{F}$ because of the symmetry of Π we obtain — as already stated — (1.233) from (1.231) by applying the Fourier transform. Note that $\mathrm{supp}(Ш\, g) \cap \mathbb{Z}$ is the empty set. For (1.234) we observe that — say for $\Phi \in C_\infty(\mathbb{R})$ — we have

$$\begin{aligned}(\Pi(D)\, Ш_N\, g, \Phi) &= (Ш_N\, g, \mathrm{sinc} * \Phi), \\ &= \sum_{n=-N}^{N} g(n)\, (\sigma_n\, \delta, \mathrm{sinc} * \Phi), \\ &= \sum_{n=-N}^{N} g(n)\, (\delta, \sigma_{-n}\, (\mathrm{sinc} * \Phi)),\end{aligned}$$

$$= \sum_{n=-N}^{N} g(n)\ (\delta,\ (\sigma_{-n}\ \mathrm{sinc}) * \Phi)),$$

$$= \sum_{n=-N}^{N} g(n)\ (\sigma_n\ \mathrm{sinc}\ ,\ \Phi).$$

Thus,

$$\Pi(D)\ Ш_N\ g = \sum_{n=-N}^{N} g(n)\ \sigma_n\ \mathrm{sinc}\ .$$

Since the left-hand side converges in $\mathcal{S}_{-3}$ so does the right-hand side and it is in this sense that (1.234) is understood.

□

Remark 1.15:

(i) *This is the well-known Shannon-Whittaker sampling theorem. It states that a function* g *with a Fourier transform having compact support in* (-1/2,1/2) *can be recovered from its sampled data, i.e.*

$$Ш\ g = \sum_{n\in\mathbb{Z}} g(n)\ \sigma_n\ \delta\ .$$

This result is trivial in the sense that it is a reformulation of the Fourier series expansion. If the coefficients $\{(e^{2\pi i n m},\ f)\}_{n\in\mathbb{Z}}$, $f = \mathcal{F}^* g$, *of the Fourier series are known —— an information which is provided by the sampling —— so is the function* f *itself and thus via a Fourier transform the function* g. *Usually it is formulated as a rescaled version of* (1.231), *which makes it less obvious. Since* sinc(x) = 0 *for* $x \in \mathbb{Z}-\{0\}$ *and* sinc(0) = 1, (1.232) *is an interpolation statement. The orthonormal expansion* (1.234) *is known as a cardinal series.*

(ii) *With reference to our previous comment about extending the definition of* Ш(D) *we note that the result of Theorem 1.9 remains valid for a larger class of generalized functions. Note that the Fourier transform of a distribution —— say in* $\mathcal{S}_{-k}$ *for some* $k \in \mathbb{N}$ *—— with support in* (-1/2,1/2) *is an entire function of polynomial growth. This indicates further possibilities of extension for Theorem 1.10. Thus, Theorem 1.10 extends in particular to functions* $g \in \mathcal{F} L_1((-1/2,1/2))$. *The* *characterization of Theorem 1.10 has merely been chosen for convenience.*

(iii) *The relationship*

$$(Ш,\Phi) = (\mathcal{F}^*Ш,\Phi) = (Ш,\mathcal{F}\Phi),$$

yields Poisson's summation formula

$$\sum_{n\in\mathbb{Z}} \Phi(n) = \sum_{n\in\mathbb{Z}} (\mathcal{F}\Phi)(n),$$

for $\Phi \in \mathcal{S}_3$.

1.1.4 On almost periodic functions

Taking the generalized eigenfunctions $(e^{2\pi i\lambda Ш})_{\lambda\in\mathbb{R}}$ of 1.1.2 , we may consider linear combinations of these as a function space with the usual linear structure. Since

$$\frac{1}{2M}\int_{-M}^{M} e^{2\pi i\lambda t}\,dt = \left\{\begin{array}{ll} 1 & \text{if } \lambda = 0 \\ \frac{\sin(2\pi\lambda M)}{2\pi\lambda M} & \text{if } \lambda \in \mathbb{R}\backslash\{0\} \end{array}\right\} \equiv \operatorname{sinc}(2\lambda M) ,$$

it is clear that $(e^{2\pi i\lambda Ш})_{\lambda\in\mathbb{R}}$ is an orthonormal basis of $\operatorname{span}((e^{2\pi i\lambda Ш})_{\lambda\in\mathbb{R}})$ w.r.t. the inner product

$$(f,g)_{\#} := \lim_{M\to\infty} \frac{1}{2M}\int_{-M}^{M} \overline{f(t)}\,g(t)\,dt , \quad f,g \in \operatorname{span}((e^{2\pi i\lambda Ш})_{\lambda\in\mathbb{R}}) . \tag{1.235}$$

Consequently, the completion $H_{\#}$ of $\operatorname{span}((e^{2\pi i\lambda Ш})_{\lambda\in\mathbb{R}})$ w.r.t. the norm $\|\bullet\|_{\#}$ induced by (1.235) is a Hilbert space

$$H_{\#} = \overline{\operatorname{span}((e^{2\pi i\lambda Ш})_{\lambda\in\mathbb{R}})}^{\|\bullet\|_{\#}} .$$

$H_{\#}$ contains the set of almost periodic functions which are defined as the uniform limits of elements in $\operatorname{span}((e^{2\pi i\lambda Ш})_{\lambda\in\mathbb{R}})$. This is is clear from the straight_forward estimate

$$\|f\|_{\#} \leq \sup_{t\in\mathbb{R}} |f(t)| \text{ for } f \in \operatorname{span}((e^{2\pi i\lambda Ш})_{\lambda\in\mathbb{R}}) . \tag{1.236}$$

the term 'almost periodic' is motivated by the fact that $e^{2\pi i\lambda \mathbf{m}}$ is periodic if λ is rational and the rational numbers are dense in $\mathbb{R}$. The space $H_\#$ is an example of a non-separable Hilbert space and its elements shall be called generalized almost periodic functions.

To have some examples of functions in $H_\#$ let us consider functions in $L_2(\mathbb{R})$ with support in a bounded interval —— say $(-N/2,N/2)$. For such functions we have

$$\mathscr{e}_{N,0}\, f \in L_2(\mathbb{R}) \text{ and } \operatorname{supp}(\mathscr{e}_{N,0}\, f) \subset (-1/2,1/2).$$

Expanding $\mathscr{e}_{N,0}\, f$ into a Fourier series yields

$$\mathscr{e}_{N,0}\, f = \sum_{n\in\mathbb{Z}} (e^{2\pi i n\mathbf{m}}, \mathscr{e}_{N,0} f)\, e^{2\pi i n\mathbf{m}} \quad \text{in } L_2((-1/2,1/2)),$$

or applying periodic extension to $\mathscr{e}_{N,0}\, f$

$$P_\#\, \mathscr{e}_{N,0}\, f = \sum_{n\in\mathbb{Z}} (e^{2\pi i n\mathbf{m}}, \mathscr{e}_{N,0} f)\, e^{2\pi i n\mathbf{m}} \quad \text{in } \mathscr{S}_{-1}.$$

Applying $\mathscr{e}_{1/N,0}$ we obtain

$$\mathscr{e}_{1/N,0}\, P_\#\, \mathscr{e}_{N,0}\, f = \frac{1}{N} \sum_{n\in\mathbb{Z}} (e^{2\pi i n\mathbf{m}/N}, f)\, e^{2\pi i n\mathbf{m}/N} \quad \text{in } \mathscr{S}_{-1}. \tag{1.237}$$

Note that $\mathscr{e}_{1/N,0}\, P_\#\, \mathscr{e}_{N,0}$ is the identity if restricted to $L_2((-N/2,N/2))$, so that (1.237) gives the periodic extension of f with period interval $(-N/2,N/2)$. From the Fourier series we get

$$\|f\| \equiv \|f\|_{L_2(\mathbb{R})} = N^{-1} \sum_{n\in\mathbb{Z}} |(e^{2\pi i n\mathbf{m}/N}, f)|^2 = N^{-1} \|\mathscr{e}_{1/N,0}\, P_\#\, \mathscr{e}_{N,0}\, f\|_\#\,,$$

which shows that

$$\mathscr{e}_{1/N,0}\, P_\#\, \mathscr{e}_{N,0}\, f \in H_\#\,.$$

The differential operator

$$D : \operatorname{span}((e^{2\pi i\lambda\mathbf{m}})_{\lambda\in\mathbb{R}}) \subset H_\# \longrightarrow H_\# \tag{1.238}$$

is densely defined and symmetric. Symmetry is obvious from

$$(D\, e^{2\pi i\lambda\mathbf{m}}, e^{2\pi i\mu\mathbf{m}}) = \lambda (e^{2\pi i\lambda\mathbf{m}}, e^{2\pi i\mu\mathbf{m}}) = \lambda \cdot \delta^\lambda_\mu = \mu\, \delta^\lambda_\mu\,, \tag{1.239}$$

where

$$\delta^\lambda_\mu = \begin{cases} 0 & \text{if } \lambda \neq \mu \\ 1 & \text{if } \lambda = \mu \end{cases}.$$

The closure of (1.238) is a selfadjoint operator $D_\#$ on $H_\#$. It is clear from the definition of $D_\#$ that

$$\sigma(D_\#) = \mathbb{R} ,$$

and the spectrum is purely a point spectrum

$$P\sigma(D_\#) = \sigma(D_\#) = \mathbb{R} . \tag{1.240}$$

The eigensolutions are in fact the elements $e^{2\pi i\lambda \boldsymbol{m}} \in H$, $\lambda \in \mathbb{R}$. The domain of $D_\#$ can be described as

$$D(D_\#) = \{ f = \sum_{n\in\mathbb{N}} c_n \exp(2\pi i\lambda_n \boldsymbol{m}) \mid \sum_{n\in\mathbb{N}} |c_n \lambda_n|^2 < \infty \text{ and } \sum_{n\in\mathbb{N}} |c_n|^2 < \infty \}.$$

As an example consider

$$\varphi = \sum_{n\in\mathbb{N}} n^{-3/4} \exp(2\pi i n^{-1/4} \boldsymbol{m}) \in H_\#.$$

We also have

$$D_\# \varphi = \sum_{n\in\mathbb{N}} n^{-1} \exp(2\pi i n^{-1/4} \boldsymbol{m}) \in H_\# ,$$

and so

$$\varphi \in D(D_\#).$$

Note that $\varphi(0)$ is not defined, which shows that we do not have a Sobolev type result.

Since any element in $H_\#$ can be written uniquely as a series converging in $H_\#$:

$$f = \sum_{\lambda\in E} \alpha_\lambda e^{2\pi i\lambda \boldsymbol{m}} , \quad E \text{ a countable subset of } \mathbb{R} ,$$

where $\alpha_\lambda \neq 0$ for $\lambda \in E$, $E \equiv \{\lambda \in \mathbb{R} \mid \alpha_\lambda \neq 0\}$, We see that by letting

$$f \longrightarrow (\alpha_{\lambda_n})_{n\in\mathbb{N}} ,$$

where $E = (\lambda_n)_{n\in\mathbb{N}}$ (in non-decreasing order) and α_{λ_n} coefficient of $\exp(2\pi i\lambda_n \boldsymbol{m})$ in the expansion of f, we have defined a unitary mapping

$$\mathcal{F}_\#: H \longrightarrow \bigoplus_{\substack{E\subset\mathbb{R} \\ E \text{ countable}}} H_E , \tag{1.241}$$

with $H_E = L_2(\mathbb{R},\delta_{\mathbb{N}}) \equiv \ell_2$, ($\delta_{\mathbb{N}}$ denotes the Dirac measure with support $\mathbb{N}$), such that

$$\mathscr{F}_{\#}\ (D_{\#}\Phi) = \lambda\ \mathscr{F}_{\#}\ (\Phi) \qquad \text{for} \quad \Phi \in D(D_{\#})\ . \tag{1.242}$$

This way we have

<u>Theorem 1.5</u>: *The operator* $D_{\#}$ *has pure point spectrum*

$$\sigma(D_{\#}) = P\sigma(D_{\#}) = \mathbb{R}\ .$$

The mapping $\mathscr{F}_{\#}$ *defined in* (1.242) *is a spectral representation of* $D_{\#}$.
The spectrum of $D_{\#}$ *is simple.*

We conclude our brief encounter with almost periodic functions with a remark about a possible extension of the presented concepts.

<u>Remark 1.16</u>: *By a slight modification of the considerations above we can introduce H-valued generalized almost periodic functions, where H is a separable Hilbert space. This is done by considering elements in the Hilbert space*

$$H_{\#}(\mathbb{R},H) := H_{\#} \otimes H.$$

If A is a selfadjoint operator in the Hilbert space H with discrete spectrum then

$$(D_{\#} \otimes A)\ v = 0,$$
$$v(0,\bullet) = v_0 \in D(A),$$

(compare Theorem A.21), where $D_{\#} \otimes A$ *is a selfadjoint operator on* $H_{\#}(\mathbb{R},H)$, *may be used as a formulation of an initial value problem. The solution is a generalized H-valued almost periodic function, since it is the limit of a trigonometric series*

$$T_N = \sum_{n=-N}^{N} \exp(-2\pi i\ \lambda_n)\ P_n\ v_0\ ,$$

where P_n *denotes the orthogonal projection on eigenspace* $N(A+\lambda_n)$, $\sigma(A) = \{\lambda_n\}_{n\in\mathbb{Z}}$.

<u>1.1.5 The higher-dimensional Fourier transform as a spectral representation</u>.

As an example of a simple higher-dimensional integral transform we shall consider in this chapter the Fourier transform in $\mathbb{R}^n$, $n \in \mathbb{N}$. The tensor product construction given below may be applied to any other integral transform in order to obtain higher dimensional integral transforms. Let D be the selfadjoint operator considered in paragraph 1.1.2 and let $(E(\lambda))_{\lambda\in\mathbb{R}}$ be its spectral family. Then

$$E_j := \overline{Id \otimes \dots \otimes Id \otimes E \otimes Id \otimes \dots \otimes Id}\ , \tag{1.243}$$

(Id denotes the identity operator, E is at position j in the tensor product), is a spectral family. The necessary properties are easily seen on decomposable elements $e_{\alpha_1} \otimes \dots \otimes e_{\alpha_n} =: e_\alpha \in \bigotimes_{i=1}^{n} L_2(\mathbb{R})\ (\equiv L_2(\mathbb{R}^n))$, $(e_i)_i$ orthonormal system in $L_2(\mathbb{R})$, and extends via linearity and closure to the spectral family $(E_j(\lambda))_{\lambda\in\mathbb{R}}$, $j=1,2,\dots,n$. The operators

$$D_j := \int_{\lambda\in\mathbb{R}} \lambda \, dE_j(\lambda)\ , \ j=1,2,\dots,n, \tag{1.244}$$

are selfadjoint extensions of the essentially selfadjoint operators

$$Id \otimes \dots \otimes Id \otimes D \otimes Id \otimes \dots \otimes Id$$

defined on the algebraic tensor product

$$\bigotimes_{i=1}^{n} C_\infty(\mathbb{R})\ .$$

The operators D_j, $j = 1,2,\dots,n$, form a set of commuting, selfadjoint operators (abelian system). This is easily seen by checking the commutativity of the resolvents on $\bigotimes_{i=1}^{n} C_\infty(\mathbb{R})$. If we write suggestively $\frac{1}{2\pi i}\nabla$ for $(D_1,\dots,D_n)$, $j = 1,2,\dots,n$, we have from the spectral theory of abelian systems

$$f((2\pi i)^{-1}\nabla)u = \int_{\mu\in\mathbb{R}^n} f(\mu)\, d\Pi(\mu)\ , \tag{1.245}$$

where $\Pi(\mu) = \prod_{j=1}^{n} E_j(\mu_j)$, $\mu = (\mu_1, \dots, \mu_n) \in \mathbb{R}^n$, for any Borel measurable function f defined on $\mathbb{R}^n$. The n-fold tensor product

$$\overline{\mathcal{F} \otimes \dots \otimes \mathcal{F}}$$

is a unitary operator from $\bigotimes_{i=1}^{n} L_2(\mathbb{R})$ onto itself. Since $\bigotimes_{i=1}^{n} L_2(\mathbb{R})$ may be identified with $L_2(\mathbb{R}^n)$ (via a unitary transformation j) we have the unitary transformation

$$\mathcal{F}_n = j \ \mathcal{F} \otimes \ldots \otimes \mathcal{F} \ j^* : L_2(\mathbb{R}^n) \longrightarrow L_2(\mathbb{R}^n) . \tag{1.246}$$

Writing $\mathcal{F}_n$ in integral form we get

$$\mathcal{F}_n(f)(\mu) = \int_{x\in\mathbb{R}^n} e^{-2\pi i\mu x} f(x)\, dx , \quad f \in L_2(\mathbb{R}^n) , \tag{1.247}$$

i.e. the <u>n-dimensional Fourier transform</u>. The integral in (1.247) has to be interpreted —— as before —— as isometric extension of $\mathcal{F}_n$ initially defined on $\bigotimes_{i=1}^{n} C_\infty(\mathbb{R})$. $\mathcal{F}_n$ is a spectral representation of $(2\pi i)^{-1} \nabla$ in the sense that $\mathcal{F}_n$ is unitary and

$$\mathcal{F}_n(D_j \Phi)(\mu) = \mu_j \mathcal{F}_n(\Phi)(\mu) , \quad \mu \in \mathbb{R}^n, \tag{1.248}$$

for all $\Phi \in D(D_j) \equiv \{\Phi \in L_2(\mathbb{R}^n) \mid \int_{\mu\in\mathbb{R}^n} |\mu_j|^2 \, d\|\Pi(\lambda)\Phi\|^2 < \infty\}$. It is also apparent that $\mathcal{F}_n$ can be regarded as a unitary mapping between tempered Sobolev spaces $\mathcal{S}_k^n = \mathcal{S}_k \otimes \ldots \otimes \mathcal{S}_k$, where the tensor product is n-fold and $k \in \mathbb{Z}$.

As a simple application of (1.245) —— (1.248), we obtain a representation of the spectral family $(E_{-\Delta}(\lambda))_{\lambda\in\mathbb{R}}$ of the Laplacian

$$(-\Delta) = 4\pi^2 \sum_{j=1}^{n} D_j^2 \text{ in } \mathbb{R}^n,$$

$$(E_{-\Delta}(\lambda)f)(\mu) = \mathcal{F}_n \chi_{K(0,2\pi\lambda)}(\mu) \mathcal{F}_n(f)(\mu) , \quad f \in L_2(\mathbb{R}^n),$$

$\chi_{K(0,2\pi\lambda)}$ denoting the characteristic function of

$$K(0,2\pi\lambda) = \{x \in \mathbb{R}^n \mid |x| \le 2\pi\lambda\} .$$

We have in particular (compare [39], p. 19)

$$\|E_{-\Delta}(\lambda)f\|^2 = \int_{2\pi|\mu|\le\lambda} |\mathcal{F}_n f(\mu)|^2 \, d\mu$$

$$= \int_0^{\lambda/2\pi} \int_{|\mu_0|=1} |\mathcal{F}_n f(\rho\cdot\mu_0)|^2 \, d\mu_0 \, \rho^2 \, d\rho \, , \quad \lambda \geq 0 \, .$$

Remark 1.17: *An interesting interaction between one-dimensional and higher-dimensional Fourier transform is featured in the Radon transform*

$$(\mathcal{R}f)(t,\omega) := \mathcal{F}^*((\mathcal{F}_n f)(\mathbb{m}\,\omega))(t), \; t \in \mathbb{R}, \; \omega \in \mathbb{R}^n, \; |\omega| = 1,$$

for $f \in C_\infty(\mathbb{R}^n)$. *Because of its composition out of Fourier transforms the Radon transform* $\mathcal{R}$ *may be extended to tempered distributions (see* [14], [22] *for details). A more familiar form of the Radon transform is obtained by evaluating the partial cancelation between* $\mathcal{F}^*$ *and* $\mathcal{F}_n$ *appearing in the definition of* $\mathcal{R}$. *This yields*

$$(\mathcal{R}\, f)(t,\omega) = (\sigma_{t\omega}\, \delta, \, f), \; t \in \mathbb{R}, \; \omega \in \mathbb{R}^n, \; |\omega| = 1,$$

for $f \in C_\infty(\mathbb{R}^n)$, *which can be written as*

$$(\mathcal{R}\, f)(t,\omega) = \int_{x \perp \omega} f(t\omega + x) \, dx \, ,$$

where the integration is over the hyperplane $\{x \in \mathbb{R}^n \mid x \perp \omega\}$.

After this excursion we come back to one-dimensional questions again. In the next chapters we shall deal with variants of the one-dimensional Fourier transform.

1.4 A selfadjoint realization of the differential expression m D + iν, ν ε IR. The Mellin transform

Remembering that the differential expression D gave rise to a selfadjoint operator it might seem surprising that it is possible to obtain a selfadjoint realization for $D + i\nu$, $\nu \in \mathbb{R}$. The solution lies in the use of a weight function in the definition of the underlying Hilbert space. We observe that $D + i\nu$, $\nu \in \mathbb{R}$, is formally symmetric with respect to the weighted L_2-inner product

$$(f,g)_\nu = \int_I e^{-4\pi\nu x}\, \overline{f(x)}\, g(x)\, dx\ . \tag{1.249}$$

where $f,g \in C_\infty(I)$ and I is the underlying interval.

Once the proper weight has been found there are two ways to develop the corresponding spectral results. One way consists in following the same procedure as for the Fourier transform. Since the necessary steps have been exemplified in chapter 1.1, we may leave this as an excercise to the reader.

In order to make use of the spectral results previously obtained for the Fourier transform we shall persue a different approach in this paragraph.

Let H_ν denote the Hilbert space generated by $C_\infty(I)$ with respect to the inner product defined in (1.249) . The mapping

$$j_\nu : H_\nu \longrightarrow L_2(I)$$

$$u \longrightarrow e^{-2\pi\nu \mathbf{m}}\, u$$

is unitary (; recall: $(e^{-2\pi\nu \mathbf{m}} u)(x) = e^{-2\pi\nu x} u(x)$, $x \in I$). Since formally

$$(D + i\nu) = e^{2\pi\nu \mathbf{m}}\, D\, e^{-2\pi\nu \mathbf{m}}\ ,$$

we see that

$$D + i\nu \text{ has a selfadjoint realization as } j_\nu^* A\, j_\nu \tag{1.250}$$

on $j_\nu^*(D(A))$ for each selfadjoint $A = d_\alpha$, $D^\#$, D and $D_\#$ of 1.1. We restrict our attention to the case $I = \mathbb{R}$. From (1.250) we already know that

$$(D + i\nu) : C_\infty(\mathbb{R}) \subset H_\nu \longrightarrow H_\nu$$

is essentially selfadjoint. Its closure - say D_ν - is unitarily equivalent to the operator D of (1.1.2). In particular,

$$j_\nu(D(D_\nu)) = D(D) .$$

This way we obtain quite easily the following theorem as an immediate consequence of the results of (1.1.2).

<u>Theorem 1.5</u>: *D_ν is selfadjoint on H_ν with domain*

$$\{\Phi \in H_\nu \mid e^{-2\pi\nu u}\, \Phi \in H_1(\mathbb{R})\},$$

and we have

$$\sigma(D_\nu) = C\,\sigma(D_\nu) = \mathbb{R} .$$

The mapping

$$\Lambda_\nu := \mathcal{F} \circ j_\nu : H_\nu \longrightarrow L_2(\mathbb{R})$$

is a spectral representation for D_ν. In particular, D_ν has spectral multiplicity 1 (i.e. has a simple spectrum).

We have taken interest in discussing D_ν since $\mathcal{F} \circ j_\nu$ is closely related to the <u>Laplace transform</u> La_ν. In fact,

$$\Lambda_\nu(f)(\lambda) = \int_{\mathbb{R}} e^{-2\pi(i\lambda x+\nu)x}\, f(x)\, dx, \tag{1.251}$$

$$=: La_\nu(f)(-2\pi(i\lambda+\nu)),\ \lambda \in \mathbb{R},$$

or

$$La_\nu(f)(p) = \Lambda_\nu(f)(i(\tfrac{p}{2\pi}+\nu)) = \int_{\mathbb{R}} e^{px}\, f(x)\, dx, \tag{1.252}$$

where p ranges on the line $i\mathbb{R}-2\pi\nu$ (see [23], p. 397).
The inverse of Λ_ν is

$$\Lambda_\nu^*(f)(x) = \int_{\mathbb{R}} e^{2\pi(i\lambda x+\nu)x}\, f(\lambda)\, d\lambda , \tag{1.253}$$

$$= \frac{1}{2\pi i} \int_{i\mathbb{R}-2\pi\nu} e^{-px} f(i(\frac{p}{2\pi}+\nu))\, dp \ , \ x \in \mathbb{R}.$$

Using (1.252) we get the inversion formula in the customary form

$$f(x) = \frac{1}{2\pi i} \int_{i\mathbb{R}-2\pi\nu} e^{-px} La_{\nu}(f)(p)\, dp \ , \ x \in \mathbb{R}. \tag{1.254}$$

The advantage of the Laplace transform compared to Fourier transform is that we can use the parameter ν to adjust the transform to an application problem (note $\mathcal{A}_0 = \mathcal{F}$). This way we may include e.g. functions with exponential growth to one side which is particularly handy in biological or economical application. The price that has to be payed for this is: exponential decay to the other side. But in connection with initial value problems this is usually not a severe restriction since we may assume that our functions are identically zero "in the past". This leads to the manner of speech of "one-sided" Laplace transform. We note for later use that the Γ-function (compare [34], p. 180) can be defined as

$$\Gamma(\alpha+1) := \mathcal{A}_{1/2\pi}(h\cdot\mathbf{m}^{\alpha})(0) \ , \ \alpha \in \mathbb{R}, \ \alpha > -1, \tag{1.255}$$

where h denotes the Heaviside function.

From (1.255) it is easily seen that

$$\Gamma(\alpha+1) = \alpha\,\Gamma(\alpha) \ , \quad \alpha > 0,$$
$$\Gamma(n+1) = n! \ , \quad n = 1, 2, 3, \ldots,$$
$$\Gamma(1) = 1 .$$

Remark 1.18:

(i) *Under suitable assumptions formulae* (1.252), (1.254) *may be extended to complex values of* λ , *but this is a special function topic beyond the scope of these lecture notes (see e.g.* [34], [32], [23]*).*

(ii) *The dependence on the parameter* ν *is usually not indicated. So, the occurrence of a parameter in the inversion formula comes as a surprise. In a sense, the considerations above explain how this parameter comes into play.*

The close relationship between the Laplace transform and the Fourier transform allows extension to tempered distributions. Since this is true for all transforms discussed in part 1 of these lecture notes we shall — rather than mentioning this possibility each time — collect comments about possible extensions of the respective integral transform in a separate section (i.e. 1.5).

1.3 A selfadjoint realization of the differential expression $D+i\nu_1+2im\nu_2$, ν_1, $\nu_2 \in \mathbb{R}$. The Gauss-Weierstrass tranform

This case can be handled analogous to the situation discussed in the last paragraph. Moreover, the situation discussed in the previous section is apparently a spacial case. The appropriate unitary mapping is

$$\begin{aligned} j_{\nu_1,\nu_2} : \quad H_{\nu_1,\nu_2} &\longrightarrow L_2(I) \\ u &\longrightarrow e^{-2\pi(\nu_1+\nu_2 m) m} u \end{aligned} \tag{1.256}$$

with H_{ν_1,ν_2} the completion of $C_\infty(I)$ with respect to the weighted norm induced by the inner product

$$(f,g)_{\nu_1,\nu_2} = (j_{\nu_1,\nu_2} f,\ j_{\nu_1,\nu_2} g), \tag{1.257}$$

for $f,g \in C_\infty(I)$. The inner product on the right-hand side of (1.257) is the usual inner product of $L_2(I)$.

Here again the particularly interesting case is $I = \mathbb{R}$ and we shall restrict our consideration to this situation. Similar as in 1.2 it can be seen that if D denotes the operator defined in 1.1.2 then

$$D_{\nu_1,\nu_2}\Phi := j^{-1}_{\nu_1,\nu_2} D\, j_{\nu_1,\nu_2} \Phi \quad \text{for } \Phi \in j^{-1}_{\nu_1,\nu_2} D(D) \subset H_{\nu_1,\nu_2},$$

is the selfadjoint extension of $D+i\nu_1+2i\nu_2 m$, ν_1, $\nu_2 \in \mathbb{R}$, defined on $C_\infty(\mathbb{R})$.

We state the immediate conclusions as

<u>Theorem 1.6</u>: D_{ν_1,ν_2} *is selfadjoint on* H_{ν_1,ν_2} *with domain*

$$\{\Phi \in H_{\nu_1,\nu_2} \mid j_{\nu_1,\nu_2}(\Phi) \in H_1(\mathbb{R})\},$$

the spectrum is

$$\sigma(D_{\nu_1,\nu_2}) = C\,\sigma(D_{\nu_1,\nu_2}) = \mathbb{R},$$

and has multiplicity 1 . *The unitary mapping*

$$\Gamma_{\nu_1,\nu_2} := \mathcal{F} \circ j_{\nu_1,\nu_2} : \quad H_{\nu_1,\nu_2} \longrightarrow L_2(\mathbb{R})$$

is a spectral representation of D_{ν_1,ν_2}.

The reason for including this operator in our investigation is that its spectral resolution is related to another well-known integral transform. The case $\nu_2 = 0$ leads back to the Laplace transform. For $\nu_2 \neq 0$ we have

$$\Gamma_{\nu_1,\nu_2}(f)(\lambda) = \int_{\mathbb{R}} \exp(-2\pi(i\lambda + \nu_1 + \nu_2 x)x)\, f(x)\, dx \tag{1.258}$$

$$= e^{2\pi\nu_2 \left[\frac{\nu_1+i\lambda}{2\nu_2}\right]^2} \int_{\mathbb{R}} e^{-2\pi\nu_2 \left[x + \frac{\nu_1+i\lambda}{2\nu_2}\right]^2} f(x)\, dx\,, \quad \lambda \in \mathbb{R}.$$

The substitution

$$\sigma = -\,2\pi\nu_2 \left[\frac{\nu_1+i\lambda}{2\,\nu_2}\right] \quad \text{and} \quad h = 2\pi\nu_2\,,$$

reveals the relationship to the Gauß-Weierstraß transform. In fact, the integral transform

$$G_{h,\nu_1}(f)(\sigma) := e^{-h\sigma^2}\, \Gamma_{\nu_1,h/2\pi}(f)(i(\pi^{-1}h\sigma+\nu_1)) \tag{1.259}$$

$$\text{for} \quad f \in H_{\nu_1,h/2\pi}\,,$$

where σ ranges over $i\mathbb{R} - \frac{\nu_1\pi}{h}$, is the <u>Gauß-Weierstraß transform</u> (see [23], p. 398) of f evaluated along this line. The dependence on a second parameter (here ν_1) is usually not indicated but disclosed by the corresponding inversion formula, (compare Remark 1.18 (ii)). The inversion formula for G_{h,ν_1} is easily obtained from $\Gamma^*_{\nu_1,\nu_2} = j^*_{\nu_1,\nu_2}\, \mathcal{F}^*$ and may be left as an exercise.

An interesting special case is $\nu_2 = -\frac{1}{2}$, $\nu_1 = 0$. For this choice of parameters we obtain a spectral representation for the tempered derivative $\mathcal{D}^*$ introduced in 1.1.2.1. Now however $\mathcal{D}^*$ is considered as a selfadjoint operator in $H_{0,-1/2}$. Analogously we can consider $\mathcal{D}$ as a selfadjoint operator in

$H_{0,1/2}$. The corresponding spectral representations are

$$\Gamma_{0,-1/2}\, \Phi = (\mathcal{F} \circ j_{0,-1/2})\, \Phi = \mathcal{F}\, (\gamma^{-1}\, \Phi) \text{ for } \Phi \in H_{0,-1/2},$$

$$\Gamma_{0,1/2}\, \Phi = (\mathcal{F} \circ j_{0,1/2})\, \Phi = \mathcal{F}\, (\gamma\, \Phi) \text{ for } \Phi \in H_{0,1/2},$$

where γ is the Gauss function

$$\gamma(x) = e^{-\pi x^2}, \quad \gamma^{-1}(x) = e^{\pi x^2}, \quad x \in \mathbb{R},$$

(see 1.1.2.1).

1.4 A selfadjoint realization of the differential expression m D + iν, ν ε IR. The Mellin transform

We shall lead the discussion of the differential expression $mD+i\nu$ in the same way as for the other differential expressions in the preceding two sections. Again we restrict our attention to the case of continuous spectrum. Let us consider $m\,D + i\nu$ on $C_\infty(\mathbb{R}^+)$. In order to obtain a formally symmetric operator we have to choose the inner product

$$(u,v)_{\nu,+} := \int_{\mathbb{R}^+} x^{-4\pi\nu-1}\,\overline{u(x)}\,v(x)\,dx\,, \qquad u,v \in C_\infty(\mathbb{R}^+)\,, \tag{1.260}$$

such that the appropriate Hilbert space is the completion of $C_\infty(\mathbb{R}^+)$ with respect to the norm $\|\bullet\|_{\nu,+}$ defined by

$$\|u\|^2_{\nu,+} := (u,u)_{\nu,+}\,, \quad u \in C_\infty(\mathbb{R}^+)\,.$$

The reason why we did not choose all of $\mathbb{R}$ as domain ($|x|^{-4\pi\nu-1}$ as weight function) is that

$$m\,D + i\nu : C_\infty(\mathbb{R}^+) \subset H_{\nu,+} \longrightarrow H_{\nu,+}$$

is already essentially selfadjoint. This can be seen directly or from the following consideration. Observing that

$$\alpha_\nu : H_{\nu,+} \longrightarrow L_2(\mathbb{R})\,,$$
$$f \longrightarrow e^{-2\pi\nu\mathbf{m}}\,f(e^{\mathbf{m}})\,,$$

is unitary, we have similar as in 1.2 , 1.3 the following theorem for

$$D_{\nu,+} := \alpha_\nu^*\,D\,\alpha_\nu\,,$$

where D is as in 1.1.2.

<u>Theorem 1.7</u>: *The operator* $D_{\nu,+}$ *is selfadjoint on* $H_{\nu,+}$ *with domain*

$$\{\Phi \in H_{\nu,+} \mid \alpha_\nu \Phi \in H_1(\mathbb{R})\}\,.$$

The spectrum of $D_{\nu,+}$ *is*

$$\sigma(D_{\nu,+}) = C\,\sigma(D_{\nu,+}) = \mathbb{R} ,$$

and has multiplicity 1 . The transform

$$\mathcal{F} \circ \alpha_\nu : H_{\nu,+} \longrightarrow L_2(\mathbb{R}) \tag{1 261}$$

is a spectral representation for $D_{\nu,+}$.

We want to write (1.261) in integral form (always keeping in mind that the integral form is defined by isometric continuation):

$$\begin{aligned} M_\nu(f)(\lambda) &:= \mathcal{F} \circ \alpha_\nu(f)(\lambda), \\ &= \mathcal{F}(e^{-2\pi\nu \mathrm{m}} f(e^{\mathrm{m}}))(\lambda), \\ &= \int_{\mathbb{R}} e^{-2\pi(i\lambda + \nu)x} f(e^x)\, dx \\ &= \int_{\mathbb{R}^+} t^{-2\pi(i\lambda+\nu)-1} f(t)\, dt \\ &=: \mathrm{Mel}_{-2\pi\nu}(f)(-2\pi(i\lambda+\nu)), \quad f \in H_{\nu,+} , \ \lambda \in \mathbb{R}. \end{aligned} \tag{1.262}$$

The transform

$$\mathrm{Mel}_{-2\pi\nu}(f)(p) = M_\nu(f)(i(\tfrac{p}{2\pi}+\nu)) , \tag{1.263}$$

$f \in H_{\nu,+}$, initially defined for p on the line $i\mathbb{R}-2\pi\nu$, is the well-known <u>Mellin transform</u> (see [23], p. 397).

The inversion formula can be calculated from

$$\begin{aligned} M_\nu^*(f)(x) &= \alpha_\nu^* \circ \mathcal{F}^*(f)(x), \\ &= \alpha_\nu^* \int_{\mathbb{R}} e^{2\pi ixt} f(t)\, dt, \\ &= \int_{\mathbb{R}} x^{2\pi(it+\nu)} f(t)\, dt, \end{aligned} \tag{1.264}$$

$$= \frac{1}{2\pi i} \int_{i\mathbb{R}-2\pi\nu} x^{-p} f(i(\tfrac{p}{2\pi}+\nu))\, dp \;,$$

for $f \in L_2(\mathbb{R})$, $x \in \mathbb{R}^+$. Equations (1.263) and (1.264) together yield the inversion formula

$$f(x) = \frac{1}{2\pi i} \int_{i\mathbb{R}+\gamma} x^{-p} \operatorname{Mel}_\gamma(f)(p)\, dp \;, \tag{1.265}$$

for $f \in H_{-\gamma/2\pi,+}$ (see [23], p. 397).

From the definition of $H_{\nu,+}$ we see an indication for the use of the Mellin transform; we can build-in power singularities in our space. In fact, the Mellin transform is used to study singularities of potential functions in corners of plane domains. The differential operator $D_{\nu,+}$ occurs when separating the two-dimensional Laplacian in polar coordinates.

In comparison with (1.255) we note that

$$\Gamma(s) = M_{-s/2\pi}(e^{-\mathbf{m}})(0) = \operatorname{Mel}_s(e^{-\mathbf{m}})(s), s > 0,$$

(see [34], p. 190).

The Mellin transform has an interesting application in the theory of the so-called Fourier integral operators. These are integral operators with kernel K of the form

$$K(x,y) = k(x\bullet y).$$

Let $\nu = -\frac{1}{4\pi}$ then $H_{\nu,+} = L_2(\mathbb{R}^+)$, $\alpha_\nu f = e^{\mathbf{m}/2} f(e^{\mathbf{m}})$ and the corresponding differential expression is $m^{1/2}\, D\, m^{1/2}$. Based on the spectral theorem we can introduce functions of $D_* := D_{-1/(4\pi),+}$ in the same way as we introduced functions of D. Functions of D —— as discussed in 1.1.3.2 —— give rise to generalizations of convolution integrals. In a similar way functions of D_* will show to be related to Fourier integral operators. Analogous to our discussion of convolutions, let us assume that for a Borel function f the operator $f(D_*)$ happens to be an integral operator

$$Kf(x) = \int_{\mathbb{R}^+} k(x,y)\, f(y)\, dy \ , \ x \in \mathbb{R}^+. \tag{1.266}$$

Since D_* and therefore $f(D_*)$ commutes with $\ell_{\alpha,0}$, $\alpha > 0$, we have for the kernel

$$k(\alpha\, x,\ y) = \alpha^{-1}\, k(x,\ \frac{y}{\alpha}),\ x,y,\alpha \in \mathbb{R}^+,$$

or with y =1 and α^{-1} renamed as y

$$k(x,y) = y^{-1}\, k(\frac{x}{y},\ 1). \tag{1.267}$$

An example is furnished by the Gegenbauer transform employed in [22] to investigate the Radon transform. Though it seems more natural to discuss functions of D_* and consequently integral operators with kernels of the type (1.267) in the framework of our approach, they do, however, not lead directly to Fourier kernels. To link up with this concept consider the unitary operator

$$\phi := M^*\, \ell_{-1,0}\, M : L_2(\mathbb{R}^+) \longrightarrow L_2(\mathbb{R}^+)\ , \tag{1.268}$$
$$u \longrightarrow \frac{1}{m}\, u(\frac{1}{m})\ .$$

An elementary calculation shows that

$$\ell_{\alpha,0}\, \phi = \phi\, \ell^*_{\alpha,0}\ . \tag{1.269}$$

Now, assume that the operator

$$\Omega := f(D_*)\ \phi\ ,$$

defined on $\phi^*\, D(f(D_*))$ happens to be an integral operator of the type described in (1.266) then using (1.269) and the above mentioned fact that

$$\ell_{\alpha,0}\, f(D_*) = f(D_*)\ \ell_{\alpha,0}\ ,\ \alpha > 0, \tag{1.270}$$

we obtain for the kernel k

$$k(\alpha x,y) = k(x,\alpha y),$$

or choosing y = 1 and renaming α as y

$$k(x,y) = k(x\ y,1). \tag{1.271}$$

Thus we have demonstrated that considering

$$f(D_*)\ \phi : \phi^*\, D(f(D_*)) \subset L_2(\mathbb{R}^+) \longrightarrow L_2(\mathbb{R}^+), \tag{1.272}$$

for any Borel function f can be regarded as a generalization of the concept of a Fourier integral operator in the same way as we considered functions of

D as generalizations of convolution integrals. Furthermore, the option remains to extend this concept to tempered distributions using the corresponding results for the Fourier transform.

As prominent examples of Fourier integral operators (see [34], chapter VIII, [23] p. 150, [13] chapters VIII, IX, XI) we mention the Hankel transform (kernel generated by $\mathbf{m}^{1/2}\ J_{\nu}$, where J_{ν} is Bessel's function of first kind of order ν) the H-transform (kernel generated by $\mathbf{m}^{1/2}\ H_{\nu}$, where H_{ν} is Struve's function of order ν) and the Y-transform (kernel generated by $\mathbf{m}^{1/2}\ Y_{\nu}$, where Y_{ν} is Bessel's function of second kind of order ν). The latter is the inverse of the of the H-transform. For $\nu > -1$ the Hankel transform is unitary, which follows from the fact that $M(\mathbf{m}^{1/2}\ J_{\nu})$ has absolute value equal to one, (see [34], p.196). Using the above notation the Hankel transform can be written by definition as

$$2^{-2\pi i D_{*}}\ \frac{\Gamma(\frac{1}{2}\ (\nu+1)\ -\ \pi i\ D_{*})}{\Gamma(\frac{1}{2}\ (\nu+1)\ +\ \pi i\ D_{*})}\ \phi \quad \text{on } L_2(\mathbb{R}^{+}). \tag{1.273}$$

The sine transform and the cosine transform on $L_2(\mathbb{R}^{+})$ (see chapter 2.2) are special cases of the Hankel transform. We shall approach the Hankel transform in a different way in chapter 2.3.

1.5 A comment on extensions of the theory. Selfadjoint realizations for general differential expression of the form p D q

As already pointed out there is an intimate relationship between the Fourier transform and those other special integral transforms we discussed in sections 1.2 — 1.4 which is due to the fact that the corresponding differential operators are unitarily equivalent. In particular, we have for $\nu, \mu, \nu_1, \nu_2 \in \mathbb{R}$

$$\mathcal{F} = \Lambda_\nu j_\nu^* = \Gamma_{\nu_1,\nu_2} j^*_{\nu_1,\nu_2} = M_\mu \alpha_\mu^* \text{ on } L_2(\mathbb{R}), \tag{1.274}$$

or

$$\Lambda_\nu = \Gamma_{\nu_1,\nu_2} (j^*_{\nu_1,\nu_2} j_\nu) = M_\mu (\alpha_\mu^* j_\nu) \text{ on } H_\nu \text{ etc. .} \tag{1.275}$$

Many classical relationships between those integral transforms can be derived by observing (1.274), (1.275). This close relationship also opens the door for transferring our results about the Fourier transform in tempered Sobolev spaces to other integral transforms. Before we look closer into this possibility we shall however consider another way to generalize our results (compare [11], p. 1502-1503).

The discussion of the Mellin transform leads to an interesting generalization. Let j be a unitary mapping from a Hilbert space H onto $L_2(\mathbb{R})$, then by extracting the general idea from previous investigations we have apparently that $\mathcal{F} \circ j$ is a spectral representation for $j^* D\, j$ defined as a selfadjoint operator on H with domain $j^* D(D)$. To relate this to differential expressions we construct H as a function space over an open interval I and define j^*, j in terms of two mapping Ω, β as

$$j^*\, u = \frac{1}{\Omega \circ \beta}\, (u \circ \beta) \text{ for } u \in L_2(\mathbb{R}), \tag{1.276}$$

and

$$j\, v := \Omega\, (v \circ \beta^{-1}) \text{ for } v \in j^*\, C_\infty(\mathbb{R} - N(\Omega)), \tag{1.277}$$

where

$\Omega: \mathbb{R} \longrightarrow \mathbb{R}$ continuously differentiable and non-negative,

such that

$N(\Omega) = \{x \in \mathbb{R} \mid \Omega(x) = 0\}$ consists only of isolated points,

and

$\beta: I \longrightarrow \mathbb{R}$ continuously differentiable with $\beta' > 0$ in I.

Then, we define $H \equiv H(j)$ as the completion of $j^* C_\infty(\mathbb{R} - N(\Omega))$ with respect to the norm $\|\bullet\|_H$ induced by the inner product determined by

$$(j^*f, j^*g)_H := (f,g) \text{ for } f,g \in C_\infty(\mathbb{R} - N(\Omega)). \tag{1.277}$$

With

$$p := \frac{1}{\beta' (\Omega\circ\beta)}, \quad q := \Omega\circ\beta,$$

we have that the following theorem holds.

<u>Theorem 1.15</u>: *The operator* $j^* D j$ *with domain* $j^* D(D) \subset H(j)$ *is a selfadjoint realization of the differential expression*

$$p\, D\, q \equiv \frac{1}{\beta' (\Omega\circ\beta)} D\, (\Omega\circ\beta) = \frac{1}{\beta'} D + \frac{1}{2\pi i} \frac{\Omega'\circ\beta}{\Omega\circ\beta}. \tag{1.266}$$

The transform

$$\mathcal{F} \circ j : H(j) \longrightarrow L_2(\mathbb{R})$$

is a spectral representation for $j^* D j$. *The spectrum of* $j^* D j$ *is*

$$\sigma(j^* D j) = C\sigma(j^* D j) = \mathbb{R},$$

and has multiplicity 1.

<u>Proof</u>: To see this result we only have to observe that j^* is in fact the inverse of j and that j^* can be extended isometrically to all of $L_2(\mathbb{R})$. This is clear, since by our assumptions $C_\infty(\mathbb{R}-N(\Omega))$ is dense in $L_2(\mathbb{R})$. Thus, j^* is unitary and j can be extended to all of H(j) by continuity as its inverse. From this the spectral result follows. Formula (1.266) is obtained by evaluating the action of $j^* D j$ on $C_\infty(\mathbb{R}-N(\Omega))$. □

We shall now indicate how the extension of the Fourier transform to tempered distributions ca be tranferred to the general class of transforms under discussion. Since j^* maps $L_2(\mathbb{R})$ onto $H(j)$ we may consider $H_k(j) = j^* \mathcal{S}_k$ as Hilbert spaces densely and continuously imbedded in $H(j)$ with inner product defined by

$$(j^*\Phi, j^*\Psi)_{H(j),k} = (\Phi,\Psi)_{\mathcal{S},k}\ , \ k \in \mathbb{N} \cup \{0\}.$$

In an analogous way to section 1.1.2.2 we can now consider bounded linear functionals on $H_k(j)$. Thus we obtain Hilbert spaces $H_{-k}(j)$, $k \in \mathbb{N}$, with the inner product defined in terms of the corresponding Riesz map and we shall use the inner product of $H(j)$ as notation

$$f(\Phi) \equiv (f,\Phi)_{H(j),0},$$

for the application of $f \in H_{-k}(j)$ to $\Phi \in H_k(j)$, $k \in \mathbb{N} \cup \{0\}$ (compare section 1.1.2.2).

By this construction we have that

$$\mathcal{F}\circ j : H_k(j) \longrightarrow \mathcal{S}_k \text{ is unitary, } k \in \mathbb{N} \cup \{0\}.$$

With the usual idea of generalizing to the dual we let for $f \in H_{-k}$

$$(j\ f, \Phi) = (f, j^*\ \Phi)_{H(j),0} \text{ for } \Phi \in \mathcal{S}_k.$$

This extends in particular $\mathcal{F}\circ j$ to spaces with negative index and we have, thus, shown the following theorem.

Theorem 1.16: *The mapping*

$$\mathcal{F}\circ j : H_k(j) \longrightarrow \mathcal{S}_k$$

is unitary for $k \in \mathbb{Z}$. *Moreover, if*

$$f(D) : \mathcal{S}_s \longrightarrow \mathcal{S}_t$$

is a bounded operator ($f(D)$ *is an extended function of* D *as discussed in 1.1.3.2) then so is*

$$f(j^*\ D\ j) = j^*\ f(D)\ j : H_s(j) \longrightarrow H_t(j),\ s,\ t \in \mathbb{Z}.$$

Remark 1.19:

(i) *We note that* $H(j)$ *is a weighted* L_2 *-space. To see this it suffices to consider* $\|\Psi\|_{H(j)}$ *for functions* Ψ *in* $j^*\ C_\infty(\mathbb{R} - N(\Omega))$:

$$\|\varphi\|^2_{H(j)} = \| j\ \varphi\|^2 = \int_{\mathbb{R}} |\Omega(t)\ (\varphi\circ\beta^{-1})(t)|^2\ dt,$$

$$= \int_{I} |\Omega(\beta(x))|^2\ |\varphi(x)|^2\ \beta'(x)\ dx.$$

As a description of H(j) *we thusly have*

$H(j) = |\Omega\circ\beta|^{-1}\ L_2(I),$

where

$$|\Omega\circ\beta|^{-1}\ L_2(I) := \{\Phi \in L_2^{loc}(I)\ |\ |\Omega\circ\beta|\ \Phi \in L_2(I)\},$$

equipped with the natural inner product

$$(|\Omega\circ\beta|\ u,\ |\Omega\circ\beta|\ v),\ u,v \in |\Omega\circ\beta|^{-1}\ L_2(I),$$

induced by the inner product $(\bullet,\bullet)$ *of* $L_2(I)$. *Recall that*

$\Phi \in L_2^{loc}(I)$ *iff* $\Phi \in L_2(J)$ *for any open interval* J *with* $\bar{J} \subset I$.

(ii) *As an application of the extension to tempered distributions let us consider the (extended) transform* M_0^* *(discussed in the previous chapter). We have e.g. on* $\mathcal{S}_{-3}$

$$M_0^* = \alpha_0^*\ \mathcal{F}^*.$$

Thus, for example

$$M_0^*\ (\text{Ш}\ f) = \sum_{n\in\mathbb{Z}} f(n)\ \mathbf{m}^{2\pi i n} \in H_{-3}(\alpha_0)\ \textit{for}\ f \in \mathcal{S}_1.$$

The expression on the right-hand side (usually written with z *instead of* $\mathbf{m}^{2\pi i n}$*) is known as the* z *transform of* Ш f *(see e.g.* [8] *p.*257 *pp).*

With this view at possible generalizations we conclude our discussion of spectral results related to first order differential expressions and turn our attention to higher-order (mainly second-order) differential expressions.

Part 2: Spectral results related to some formally selfadjoint differential expressions of second order

In this second part of our lecture notes we shall investigate the spectral properties of some second order differential expressions.

We shall be dealing with differential expressions of the type

$$\mathfrak{a} = \mathfrak{b}^* \mathfrak{b}, \tag{2.1}$$

where $\mathfrak{b}$ is a formal differential operator of the type discussed in 1.5 and $\mathfrak{b}^*$ is its formal adjoint with respect to a weighted L_2 -inner product. Operators of this type can be considered as coming from studying certain first order systems of differential operators. Indeed, squaring the formally selfadjoint 2×2 operator matrix

$$\begin{bmatrix} 0 & \mathfrak{b} \\ \mathfrak{b}^* & 0 \end{bmatrix}, \tag{2.2}$$

we obtain the diagonal matrix

$$\begin{bmatrix} \mathfrak{b}\,\mathfrak{b}^* & 0 \\ 0 & \mathfrak{b}^*\,\mathfrak{b} \end{bmatrix} = \begin{bmatrix} 0 & \mathfrak{b} \\ \mathfrak{b}^* & 0 \end{bmatrix} \begin{bmatrix} 0 & \mathfrak{b} \\ \mathfrak{b}^* & 0 \end{bmatrix}. \tag{2.3}$$

There are many possibilities to make this into a selfadjoint operator matrix. We shall only consider the specific case where the matrix is of the form

$$\begin{bmatrix} 0 & B \\ B^* & 0 \end{bmatrix}. \tag{2.4}$$

Here B denotes a densely defined, closed linear operator from some Hilbert space into another based on the formal differential operator matrix in (2.2). In terms of the scalar differential expression (2.1) this means we consider selfadjoint realizations of $\mathfrak{a}$ of the form B^*B (which corresponds to the second diagonal entry in (2.3)). It actually has many advantages to consider systems rather than the second order equation. Their respective spectral properties are equivalent. For sake of comparability with existing spectral statements, however, we shall adopt the scalar point of view.

To be more specific let

$$\ell = q\,\rho(2\pi D)\rho^{-1}, \tag{2.5}$$

where q, ρ are non-negative real valued functions. Here we include the scaling factor 2π in order to link up with notations used in the literature. If we choose Hilbert space $p^{1/2}L_2(I)$ (over the underlying interval I, p another non-negative real-valued function) as inital space of ℓ and Hilbert space $q^{1/2}L_2(I)$ as terminal space and assuming that the non-regular points of q^{-1}, q, p, p^{-1}, ρ, ρ^{-1} are a set E of isolated points then by choosing functions in $C_\infty(I-E)$ we see that ℓ with this domain is a densely defined operator from $p^{1/2}\,L_2(I)$ into $q^{1/2}\,L_2(I)$ and ℓ^* is a densely defined operator from $q^{1/2}\,L_2(I)$ into $p^{1/2}\,L_2(I)$ (see remark 1.19 for the notation). We shall have non-regular points only on the boundary of the interval, so we can actually choose functions in $C_\infty(I)$ to establish that ℓ is densely defined with domain $C_\infty(I)$. With this domain ℓ, $\ell^* = p\,\rho^{-1}\,(2\pi D)\,\rho$ are closable, let B, C be their respective closures then

$$B\colon D(B) \subset p^{1/2}\,L_2(I) \longrightarrow q^{1/2}\,L_2(I), \tag{2.6}$$

$$C\colon D(C) \subset q^{1/2}\,L_2(I) \longrightarrow p^{1/2}\,L_2(I), \tag{2.7}$$

are closed operators with domains D(B), D(C), respectively. The domain D(B) is a Hilbert space in fact the closure of $C_\infty(I)$ with respect to the graph norm of B. The graph norm of B is induced by the inner product

$$(q^{-1/2}\ell u, q^{-1/2}\ell v) + (p^{-1/2}u, p^{-1/2}v),$$

for u, v $\in C_\infty(I)$, (the inner product $(\bullet,\bullet)$ is the inner product of $L_2(I)$). Similarly, D(C) is a Hilbert space with inner product

$$(p^{-1/2}\ell^* u, p^{-1/2}\ell^* v) + (q^{-1/2}u, q^{-1/2}v),\quad u,v \in D(C).$$

To recover well-known results from the theory of special functions and integral transforms we shall be concerned mainly with two selfadjoint extensions of the closure A of a defined on $C_\infty(I)$. These are

$$A_D := B^* B, \quad (2.8)$$

and

$$A_N := C C^*, \quad (2.9)$$

sometimes referred to as Dirichlet and Neumann case. This is due to the fact that being in D(B) or D(C) usually induces a boundary condition, so that elements in $D(A_D)$, $D(A_N)$ usually have to satisfy typical boundary conditions referred to as Dirichlet and Neumann boundary conditions, respectively.

In general, we have that D(B) can be regarded as a subspace of $D(C^*)$ and correspondingly D(C) as a subspace of $D(B^*)$.In many instances, however, we shall find that $D(B) = D(C^*)$ i.e. $B = C^*$. Observing that $D(A_D)$ dense in D(B) and that $D(A_N)$ dense in $D(C^*)$ we have

$$A_D = A_N \text{ iff } B = C^*. \quad (2.10)$$

In particular, if A is selfadjoint we have

$$A = A_D = A_n.$$

In the examples presented in the following sections we shall use — occasionally with other indices — lower indices D and N and index as generic notations to indicate the respective cases. We shall also use the letters A, B, C, $\mathcal{a}$, $\mathcal{b}$ (with indices) to refer generically to respective operators constructed as described in the above.

Let us recall the concept of a Green's function from the solution theory of ordinary differential equations. Green's function is a family of solutions $G_\lambda(\bullet,y)$, $y \in I$, $\lambda \in \mathbb{C}$, of

$$(\mathcal{a} - \lambda)\, G_\lambda(\bullet,y) = p(y)\, \sigma_y \delta. \quad (2.11)$$

For our purposes equation (2.11) shall be understood in the weak sense of the underlying Hilbert space $H = p^{1/2} L_2(I)$ with inner product $(\bullet,\bullet)_H$. If λ is in the resolvent set $\rho(A)$ of any of those extensions A discussed above then the solution v of the inhomogeneous problem

$$(\mathcal{a} - \lambda)\, v = f,$$

where f, v are elements of the underlying Hilbert space H, can be expressed as

$$v(y) = (G_\lambda(\bullet,y),\ f)_H,$$

provided that $G_\lambda(\bullet,y)$ is in D(B) in the Dirichlet case and in $D(C^*)$ in the Neumann case. The first case usually induces boundary conditions for Green's function $G_\lambda(\bullet,y)$, $y \in I$. We can reformulate (2.11) in the Dirichlet case as

$$(q^{-1/2}\ell G_\lambda(\bullet,y),\ q^{-1/2}\ell\varphi) - \bar{\lambda}\ (p^{-1/2}G_\lambda(\bullet,y),\ p^{-1/2}\varphi) = \varphi(y), \tag{2.12}$$

for all $\varphi \in D(B)$. In the Neumann case we have to take $\varphi \in D(C^*)$. Thus the existence of Green's function becomes a matter of applying the Lax-Milgram theorem. Green's function as a solution of (2.12) or (2.13) (or (2.11)) has the form

$$G_\lambda(x,y) = c \begin{cases} u_\lambda(x)\ v_\lambda(y) & \text{for } x \geq y \\ u_\lambda(y)\ v_\lambda(x) & \text{for } x \leq y \end{cases},\quad x,y \in I, \tag{2.13}$$

where u_λ, v_λ are suitably chosen homogeneous solutions of (2.11) (i.e. (2.11) with $f \equiv 0$). The constant c appearing in (2.13) has to be adjusted in order to provide the proper behaviour at x = y. We can write (2.13) as

$$G_\lambda(x,y) = c\ (h(x-y)\ u_\lambda(x)\ v_\lambda(y) + h(y-x)\ u_\lambda(y)\ v_\lambda(x)),\quad x,y \in I. \tag{2.14}$$

Recalling that $2\pi i\ D\ h = \delta$ (h Heaviside function) we see that (assuming sufficient regularity of the coefficients and solutions u,v in I)

$$(\mathfrak{a} - \lambda)\ G_\lambda(.,y) = -\ c\ (p\ \sigma_y\delta\ q\ u'_\lambda\ v_\lambda(y) - p\ \sigma_y\delta\ q\ u_\lambda(y)\ v'_\lambda),$$

$$= c\ p(y)\ q(y)\ (u_\lambda(y)\ v'_\lambda(y) - u'_\lambda(y)\ v_\lambda(y))\ \sigma_y\delta.$$

Thus, in order to satisfy (2.11) we have to have

$$c = \{q(y)\ W(u_\lambda,v_\lambda)(y)\}^{-1},\quad y \in I,$$

where

$$W(u_\lambda,v_\lambda) := \det \begin{Bmatrix} u_\lambda & v_\lambda \\ u'_\lambda & v'_\lambda \end{Bmatrix},$$

is the Wronskian determinant of u_λ, v_λ . An elementary calculation shows that $(q\ W)' = 0$, i.e.

$$q\ W = \varepsilon_0 \text{ for some constant } \varepsilon_0 \in \mathbb{R}.$$

This implies that

$$c = \varepsilon_0^{-1}.$$

Also note for later reference that by a similarly elementary calculation we have

$$p\, W(\ell\, u_\lambda, \ell\, v_\lambda) = \lambda\, q\, W(u_\lambda, v_\lambda). \tag{2.15}$$

With this construction we can represent the resolvent $(A - \lambda)^{-1}$ of A as

$$(A - \lambda)^{-1} f = (G_\lambda(\cdot, y), f)_H , \tag{2.16}$$

for say $f \in C_\infty(I)$. On the other hand, if we can show that the integral operator

$$(R_\lambda f)(y) = (G_\lambda(\cdot,y), f)_H , \; y \in I, \; f \in C_\infty(I)$$

converges uniformly as $\lambda \in \rho(A)$ tends to $\lambda_0 \in \mathbb{C}$, i.e.

$$R_\lambda \longrightarrow R \text{ with respect to the operator norm,}$$

for some bounded operator R as $\lambda \longrightarrow \lambda_0$, then $\lambda_0 \in \rho(A)$ and

$$(A - \lambda_0)^{-1} = R.$$

With these paraphrases of the theory of ordinary differential equations we are prepared to approach the problems of the second part of these lecture notes.

2.1 Some polynomial eigensolutions

In this paragraph we shall consider the so-called classical polynomials. Since we are here in the spectral theoretically less interesting case of discrete spectrum, we will only give an overview of the corresponding operators, basic spectral properties together with some comments and hints for proofs. The ideas how to prove the results have been laid out in the above. Necessary special function material may be taken from [23], chapters IV and V. To have a guideline how to determine the spectral properties of the respective differential operators we shall consider the first example in somewhat more detail. For our purpose it is sufficient to consider the Neumann case.

2.1.1 Legendre polynomials.

Differential expression: $\mathfrak{a} := 2\pi D\ (1-m^2)\ 2\pi D$.

Selfadjoint realization:

Selfadjoint extension $A = A_N = C\ C^*$ of A . The corresponding differential operator C is

$$C : D(C) \subset (1-m^2)^{1/2}\ L_2((-1,1)) \longrightarrow L_2((-1,1)),$$

with

$$C\ \varphi = 2\pi\ D\ \varphi \equiv \mathfrak{c}^*\ \varphi,\ \varphi \in C_\infty((-1,1)),$$

and $C_\infty((-1,1))$ dense in $D(C)$ with respect to the graph norm. We also have

$$\mathfrak{c} = 2\pi\ (1-m^2)\ D.$$

Domain:

$$D(A) = \{w \in L_2((-1,+1))\ |\ (1-m^2)^{1/2}\ w' \in L_2((-1,+1)),\ \mathfrak{a}w \in L_2((-1,+1))\}.$$

Comment: By definition $D(A) = \{u \in D(C^*) \mid C^*u \in D(C)\}$. To see that D(A) has the description given above it suffices to show that $C_\infty((-1,1))$ is dense in

$$D(C^*) = \{w \in L_2((-1,+1)) \mid (1-m^2)^{1/2} w' \in L_2((-1,+1))\}.$$

Then we have in the notation of the introductory part of chapter 2

$$C^* = B,$$

which implies not only the above description of D(A) but also the selfadjointness of A (see (2.10)).

To see that $C_\infty((-1,1))$ is dense in $D(C^*)$ we calculate the ortho-complement of $C_\infty((-1,1))$ with respect to the graph inner product of C^*. An element in the ortho-complement of $C_\infty((-1,1))$ has to be a solution of the differential equation

$$(\mathfrak{a} + 1)\ u = 0.$$

This can be confirmed by the technique used in the proof of Lemma 1.1. Candidates for solutions are P_ν , Q_ν for $\nu = \frac{1}{2}(5^{1/2}-1)$. From the asymptotic behaviour of these (see below), we can conclude that neither is in $D(C^*)$. Thus, we have indeed $D(B) = D(C^*)$.

Spectrum: $\sigma(A) = P\sigma(A) = \langle t \in \mathbb{N} \mid t = n(n+1)\ ,\ n \in \mathbb{N} \cup \{0\}\rangle$, (recall that $\mathbb{N}$ denotes the set of positive integers 1,2,...).

Comment: From the regularity result Lemma 1.1 it follows that the classical solutions are the only solutions of the homogeneous system. Consequently there are two candidates for eigensolutions:

$$P_\nu\ ,\ Q_\nu\ ,\ \nu \in \mathbb{R},\ \lambda = \nu(\nu+1) \in \mathbb{R}^+,$$

where $\nu > -1/2$ without loss of generality. We have

$$P_\nu(x) \sim 1 \text{ as } x \longrightarrow 1,$$

$$Q_\nu(x) \sim \ln(1-x) \text{ as } x \longrightarrow 1, \tag{2.17}$$

and

$$P_\nu(x) \sim \ln(1+x) \text{ as } x \longrightarrow -1,\ \nu \notin \mathbb{N} \cup \{0\},$$

$$Q_\nu(x) \sim \ln(1+x) \text{ as } x \longrightarrow -1,\ \nu \notin -\tfrac{1}{2} + \mathbb{N}, \tag{2.18}$$

here the symbol "~" is used as a short-hand notation for "asymptotically proportional with a non-zero factor". We shall employ this notation when we do not particularly care for the precise factor. We observe that by the relation

$$(1-m^2)\ Q'_\nu = -\nu\ m\ P_\nu + \nu\ P_{\nu-1},$$

(see [23], p. 171) and the asymptotic behaviour of $P_\nu(x)$, $\nu \in \mathbb{R}$, as $x \longrightarrow 1$ ([23], p. 196) we have

$$(1-m^2)^{1/2}\ Q'_\nu \notin L_2((-1,1)). \tag{2.19}$$

This leaves P_ν as a possible candidate. Since

$$\ell_{-1,0}\ P_\nu = \cos(\pi\nu)\ P_\nu - 2\ \pi^{-1}\sin(\pi\nu)\ Q_\nu, \tag{2.20}$$

the asymptotic behaviour of P_ν, Q_ν at 1 implies the asymptotic behaviour of P_ν at -1. In particular we have

$$(1-m^2)^{1/2}\ P'_\nu \notin L_2((-1,1)) \text{ for } \nu \notin \mathbb{Z}. \tag{2.21}$$

This leaves P_n as possible eigensolutions, $n \in \mathbb{N} \cup \{0\}$ (see [23], p.170). The P_n are polynomials ([23], p. 174) and thus certainly in $D(C^*)$ and consequently in $D(A)$. The functions P_n, $n \in \mathbb{N} \cup \{0\}$, are obviously the only eigensolutions of A. The rest of the real axis is part of the resolvent set. We observe that Green's function is — according to our general account — given by

$$G_\lambda(x,y) = c\ (h(x-y)\ u_\lambda(x)\ v_\lambda(y) + h(y-x)\ u_\lambda(y)\ v_\lambda(x)), \tag{2.22}$$

where $u_\lambda = P_\nu$, $v_\lambda = Q_\nu - \frac{\pi}{2}\cot(\pi\nu)\ P_\nu$, $\lambda = \nu\ (\nu+1) \in \mathbb{C}-\mathbb{R}$, and $c = 1$ (compare [23], p. 170). Note in particular that $G_\lambda(\bullet,y)$, $y \in (-1,1)$, is indeed in $D(C^*)$ for $\lambda \neq n(n+1)$ for all $n \in \mathbb{N} \cup \{0\}$. Since P_ν, Q_ν are continuous with respect to ν (even differentiable see [23], p.178) there is no problem in passing to the limit $\lambda_0 = \nu_0\ (\nu_0+1)$, $\nu_0 \notin \mathbb{N}$. For the kernels $G_\lambda(\bullet,y)$ we have

$$\left| \int_{(-1,1)} \overline{G_\lambda(x,z)}\, G_\lambda(y,z)\, dz \right| \le C, \tag{2.23}$$

uniformly in $x,y \in (-1,1)$ and $\lambda = \nu(\nu+1)$ for ν in a compact set not containing points in $\mathbb{N} \cup \{0\}$ and some constant C. In fact the kernels themselves are uniformly bounded. Thus, by Fubini's theorem we have that

$$\| R_\lambda f \| \le C^{1/2} \|f\| \text{ for } f \in C_\infty((-1,1)).$$

This shows in particular that

$$\rho(A) = \mathbb{C} - \{\lambda \in \mathbb{R} \mid \lambda = n(n+1),\ n \in \mathbb{N} \cup \{0\}\}. \tag{2.24}$$

Eigensolution: $N(A-\lambda) = \mathrm{span}\{(n+\frac{1}{2})^{1/2} P_n\}$, $\lambda = n(n+1)$, $n \in \mathbb{N}$, P_n, $n \in \mathbb{N}$, are the so-called Legendre polynomials. The polynomials

$$((n+\tfrac{1}{2})^{1/2} P_n)_{n \in \mathbb{N}}$$

are an orthonormal Hilbert basis of $L_2((-1,+1))$.

2.1.2 Gegenbauer polynomials

Differential expression: $a := (1-m^2)^{-\nu+\frac{1}{2}}\, 2\pi D\, (1-m^2)^{\nu+\frac{1}{2}}\, 2\pi D$.

Selfadjoint realization:

Extension $A_{N,\nu} = C_\nu C_\nu^*$ of A_ν (the closure of a defined on the set $C_\infty((-1,+1))$) in Hilbert space $H_\nu = (1-m^2)^{-\nu/2+1/4} L_2((-1,1))$ with the induced norm defined by

$$\|\Phi\|_H^2 = \int_{(-1,+1)} (1-x^2)^{\nu-\frac{1}{2}} |\Phi(x)|^2\, dx ,$$

for $\Phi \in C_\infty((-1,+1))$, $\nu \in \mathbb{R}$. The operator C_ν is a closed operator from

$$(1-m^2)^{\nu/2+1/4} L_2((-1,1)) \text{ into } (1-m^2)^{-\nu/2+1/4} L_2((-1,1))$$

defined by

$$C_\nu \varphi = c_\nu^* \varphi \equiv (1-m^2)^{-\nu+\frac{1}{2}}\, 2\pi D\, \varphi,\ \varphi \in C_\infty((-1,1)).$$

The adjoint operator C^* corresponds to the formal differential expression

$$\mathscr{C}_\nu = (1-m^2)^{\nu+\frac{1}{2}} 2\pi D.$$

To simplify the discussion we shall assume that $\nu > -1/2$.

Domain:

$D(A_{N,\nu}) = \{\Phi \in H \mid (1-m^2)^{1/2} \Phi' \in H,\ a\Phi \in H\}$ for $\nu \notin (-1/2,1/2)$,

$D(A_{N,\nu}) = \{\Phi \in H \mid (1-m^2)^{1/2}\Phi' \in H,\ a\Phi \in H$, and

$$(1-x^2)^{(\nu+1/2)/2} \Phi'(x) \longrightarrow 0 \text{ as } x \longrightarrow \pm 1 \},$$

for $\nu \in (-1/2,1/2)$.

Comment: As solutions of

$$(a - \lambda) u = 0,$$

we have

$$c_\alpha^\nu := (1-m^2)^{-\nu/2+1/4} P_{\alpha+\nu-1/2}^{-\nu+1/2},$$

and

$$d_\alpha^\nu := (1- m^2)^{-\nu/2+1/4} Q_{\alpha+\nu-1/2}^{-\nu+1/2},$$

(compare [12], p 179), with $\lambda = \alpha(\alpha+2\nu) = (\alpha + \nu)^2 - \nu^2$. We may choose without loss of generality

$$\alpha + \nu > 0 \text{ and } \alpha \neq 0 \text{ if } \lambda \in \mathbb{R}^+,$$

$$\alpha > 0 \text{ if } \lambda \in \mathbb{R}^+ \text{ and } \nu \in (-1/2,1/2),$$

$$\alpha = 0 \text{ if } \lambda = 0 ,$$

and also $\nu \neq 1/2$, since this case has been considered in the previous section. If $0 > \lambda \geq -\nu^2$ choose the largest root. In all other cases we choose the root α with positive imaginary part. We can employ the asymptotic properties of the Legendre functions to show the above description of $D(A_{N,\nu})$ similarly to the considerations for the Legendre polynomials. The asymptotic behaviour of P_ν^μ, Q_ν^μ at the boundary point +1 is given by

$$\Gamma(\nu-\mu+1)\ Q_\nu^\mu(x) \sim \Gamma(\nu+\mu+1)\ (1-x)^{\mu/2} \text{ as } x \longrightarrow 1, \tag{2.25}$$

for $\mu \in \mathbb{R}^-$, and

$$P_\nu^\mu(x)\ (1-x)^{\mu/2} \sim 1 \quad \text{as } x \longrightarrow 1,\ \mu < 1, \tag{2.26}$$

(see [23], p. 197). To obtain information about the asymptotic behaviour of Q_ν^μ at 1 for $\mu \in \mathbb{R}$ we observe that the case $\mu = 0$ has already been dealt with in the previous paragraph and that

$$\begin{aligned} \Gamma(\nu+\mu+1)\ Q_\nu^{-\mu} &= \Gamma(\nu-\mu+1)\ [\cos(\pi\mu)\ Q_\nu^\mu + \tfrac{\pi}{2}\sin(\pi\mu)\ P_\nu^\mu\ ], \\ \Gamma(\nu+\mu+1)\ P_\nu^{-\mu} &= \Gamma(\nu-\mu+1)\ [\cos(\pi\mu)\ P_\nu^\mu - \tfrac{2}{\pi}\sin(\pi\mu)\ Q_\nu^\mu\ ]. \end{aligned} \tag{2.27}$$

The asymptotic behaviour of P_ν^μ, Q_ν^μ at (-1) can be deduced from

$$\begin{aligned} \ell_{-1,0}\ P_\nu^\mu &= \cos(\pi(\nu+\mu))\ P_\nu^\mu - \tfrac{2}{\pi}\sin(\pi(\nu+\mu))\ Q_\nu^\mu\ , \\ \ell_{-1,0}\ Q_\nu^\mu &= -\cos(\pi(\nu+\mu))\ Q_\nu^\mu - \tfrac{\pi}{2}\sin(\pi(\nu+\mu))\ P_\nu^\mu\ . \end{aligned} \tag{2.28}$$

This implies

$$\begin{aligned} &c_\alpha^\nu(x) \sim 1 \quad \text{as } x \longrightarrow 1, \\ &c_\alpha^\nu(x) = (-1)^\alpha\ c_\alpha^\nu(-x) \quad \text{for } \alpha \in \mathbb{Z}, \\ &c_\alpha^\nu(x)/d_\alpha^\nu(x) \sim 1 \text{ as } x \longrightarrow -1 \text{ for } \alpha \notin \mathbb{Z},\ \nu > 1/2. \end{aligned} \tag{2.29}$$

For $\nu \in (-1/2,1/2)$ we have

$$c_\alpha^\nu(x) \sim \sin(\pi(\alpha+\nu-1/2))\ , \tag{2.30}$$

as $x \longrightarrow -1$.

For the other solution d_α^ν we get

$$d_\alpha^\nu(x)\ (1-x)^{\nu-1/2} \sim 1 \quad \text{as } x \longrightarrow 1,\ \nu > 1/2,$$

and

$$d_\alpha^\nu(x)\ (1-x^2)^{\nu-1/2} \sim \cos(\pi\alpha) \quad \text{as } x \longrightarrow -1, \tag{2.31}$$

where the latter holds for $\alpha + 2\nu \neq 0$, $-\alpha \notin \mathbb{N}$, $\nu > 1/2$.
For $\nu \in (-1/2,1/2)$ we have

$$\begin{aligned} &d_\alpha^\nu(x) \sim 1 \quad \text{as } x \longrightarrow 1 \\ &d_\alpha^\nu(x) \sim \cos(\pi(\alpha+\nu-1/2)) \quad \text{as } x \longrightarrow -1 \end{aligned} \tag{2.32}$$

To characterize $D(B_\nu)$ we need information about the asymptotic behaviour of the derivatives of c_α^ν, d_α^ν. This information is provided by the relations

$$(c_\alpha^\nu)' = \lambda\, c_{\alpha-1}^{\nu+1}, \quad (d_\alpha^\nu)' = \lambda\, d_{\alpha-1}^{\nu+1}.$$

Thus, we obtain for $\nu > -1/2$

$$(c_\alpha^\nu)' \sim \lambda \text{ as } x \longrightarrow 1,$$

$$(c_\alpha^\nu)' \sim 1 \text{ or } (c_\alpha^\nu)' \sim 0 \text{ as } x \longrightarrow -1, \ \alpha \in \mathbb{N} \cup \{0\},$$

$$(c_\alpha^\nu)' \ (1+x)^{(\nu+1/2)} \sim \lambda \quad \text{as } x \longrightarrow -1, \ \alpha \notin \mathbb{Z}, \ \alpha + 2\nu \neq -1. \tag{2.33}$$

$$(d_\alpha^\nu)' \sim \lambda\ (1+x)^{-(\nu+1/2)} \quad \text{as } x \longrightarrow \pm 1, \ \alpha + 2\nu \neq -1.$$

The boundary condition follows from the Sobolev type continuity estimate

$$|\varphi(s) - \varphi(t)|^2 \le$$

$$\le \int_{(s,t)} (1-x^2)^{\nu-1/2} dx \int_{(s,t)} (1-x^2)^{\nu-1/2} \ |(\mathcal{C}^* \varphi)(x)|^2 \ dx, \tag{2.34}$$

for $\varphi \in D(B_\nu^*)$, $1 > t \ge s > 0$. Since $C_\infty((-1,1))$ is dense in Hilbert space $D(C_\nu) \subset D(B_\nu^*)$ we have in particular that

$$(C_\nu^* \ u)\restriction_{\pm 1} = 0,$$

for $u \in D(A_{N,\nu})$. We have in fact from (2.34) the stronger property that for $v \in D(C_\nu)$

$$(1-x^2)^{-\nu/2-1/4} \ v(x) \longrightarrow 0 \quad \text{as } x \longrightarrow \pm 1 \mp.$$

Noting that

$$(1-m^2)^{-\nu/2-1/4} \ \mathcal{C}\ u = -i\ (1-m^2)^{\nu/2+1/4} \ u',$$

the boundary condition for $u \in D(A_{N,\nu})$ thus becomes

$$(1-x^2)^{\nu/2+1/4} \ u'(x) \longrightarrow 0 \text{ as } x \longrightarrow \pm 1 \mp.$$

On the other hand, an element u in the ortho-complement of $D(C_\nu)$ in

$$\{\Phi \in D(B_\nu^*) \mid (1-x^2)^{-(\nu+1/2)/2} \ \Phi(x) \longrightarrow 0 \text{ as } x \longrightarrow \pm 1 \mp \},$$

has to be a solution of

$$(\ell_\nu \ell_\nu^* + 1)\ u = 0.$$

Candidates for solutions are

$$\ell\ c_\alpha^\nu = -i\ \lambda\ (1-m^2)^{\nu+1/2}\ c_{\alpha-1}^{\nu+1},\quad \ell\ d_\alpha^\nu = -i\ \lambda\ (1-m^2)^{\nu+1/2}\ d_{\alpha-1}^{\nu+1},$$

(note that by (2.15) we have linear independence).

From the asymptotic properties of c_α^ν, d_α^ν it is clear that there is no such solution in $D(B_\nu^*)$ satisfying the boundary condition and therefore we have shown

$$D(C_\nu) = \{\Phi \in D(B_\nu^*)\ \mid\ (1-x^2)^{-(\nu+1/2)/2}\ \Phi(x) \longrightarrow 0 \text{ as } x \longrightarrow \pm 1\mp\ \},$$

and consequently the characterization given for $D(A_{N,\nu})$. By a similar reasoning we see that for $\nu \in [1/2,\infty)$ we have

$$B_\nu = C_\nu^*,$$

and thus the characterization $D(A_{N,\nu})$ in this case.

Spectrum: $\sigma(A_{N,\nu}) = P\sigma(A_{N,\nu}) = \{t \in \mathbb{R}\ \mid\ t = n(n+2\nu)\ ,\quad n \in \mathbb{N} \cup \{0\}\}$, for $\nu > -1/2$.

Comment: That for $\nu > 1/2$ the point spectrum is given by

$$\{t \in \mathbb{R}\ \mid\ t = n(n+2\nu)\ ,\ n \in \mathbb{N} \cup \{0\}\}$$

is clear from the discussion of the asymptotic behaviour of c_α^ν , d_α^ν . The eigensolutions c_n^ν , $n \in \mathbb{N} \cup \{0\}$, are polynomials. To see that all other point belong to the resolvent set we have to consider only $\lambda \in \mathbb{R}^+$, $\lambda \notin P\sigma(A_{N,\nu})$, since all other points are clearly either point spectrum or resolvent set. We can reason as in the previous paragraph to establish that $\mathbb{R}^+ - P\sigma(A_{N,\nu}) \subset \rho(A_{N,\nu})$. Green's function is given by

$$G_\lambda(x,y) = c\ (h(x-y)\ u_\lambda(x)\ v_\lambda(y) + h(y-x)\ u_\lambda(y)\ v_\lambda(x)), \tag{2.35}$$

with $u_\lambda = c_\alpha^\nu$ and $v_\lambda = \sin(\pi\alpha)\ d_\alpha^\nu - \frac{\pi}{2} \cos(\pi\alpha)\ c_\alpha^\nu$ and

$$c = \frac{\Gamma(\alpha+2\nu)}{\Gamma(\alpha+1)\ \sin(\pi\alpha)}\ ,$$

for $x,y \in (-1,1)$, $-\alpha \notin \mathbb{N}$, $-\alpha-2\nu \notin \mathbb{N} \cup \{0\}$. In the calculation we have made use of (2.13) and the formula for the Wronskian of P^{μ}_{ν}, Q^{μ}_{ν} given in [23] p. 170. That

$$\rho(A_{N,\nu}) = \mathbb{C} - P\sigma(A_{N,\nu}),$$

now follows analogously to the considerations in 2.1.1. Note that isolated points not belonging to $P\sigma(A_{N,\nu})$ that have to be excluded because of problems with the validity of the asymptotic results do not have to be studied separately. As a consequence of the spectral theorem they would have to be point spectrum which contradicts the assumption. The case $\nu \in (-1/2,1/2)$ can be dealt with similarly. Here of course we have to satisfy the boundary condition.

Eigensolutions:

$$N(A_{N,\nu}-\lambda) = \mathrm{span}\{[\frac{(n+\nu)\ \Gamma(n+2\nu)}{n!}]^{1/2}\ c^{\nu}_{n}\}\ ,\quad \lambda = n(n+2\nu)\ ,\ n \in \mathbb{N} \cup \{0\},$$

$\nu > -1/2$. The polynomials

$$C^{\nu}_{n} = (-1)^{1/4-\nu/2}\ \pi^{1/2}\ 2^{1/2-\nu}\ (\Gamma(n+2\nu)\ n!)^{-1}\ c^{\nu}_{n}, \tag{2.36}$$

are the so-called Gegenbauer polynomials, $n \in \mathbb{N}$, $\nu > -1/2$. The set $\{[\frac{(n+\nu)\ \Gamma(n+2\nu)}{n!}]^{1/2}\ c^{\nu}_{n}\}_{n\in\mathbb{N}}$ is an orthonormal Hilbert basis for H_{ν}.
For $\nu > 1/2$ we have of course

$$N(A_{N,\nu}-\lambda) = N(A_{\nu}-\lambda).$$

Comment: That the eigenfunctions are of the form stated follows from the asymptotic properties of c^{ν}_{α}, d^{ν}_{α}.

Remark 2.1: *The above differential expression is closely related to the one for Legendre functions (compare 2.7)*

$$\tilde{a}_{\nu} = 2\pi D\ (1-m^{2})\ 2\pi D\ +\ (\nu-1/2)^{2}\ (1-m^{2})^{-1} + 1/4.$$

In fact we have a unitary equivalence k_{ν} *between them*

$$k_{\nu}\ f = (1-m^{2})^{(\nu-1/2)/2}\ f,\ f \in H_{\nu},$$
$$k_{\nu}: H_{\nu} \longrightarrow L_{2}((-1,1)).$$

Thus, we obtain $k_{\nu}\ (A_{N,\nu}+\nu^{2})\ k^{*}_{\nu} = \tilde{A}_{N,\nu}$, *where* $\tilde{A}_{N,\nu}$ *denotes the corresponding Neumann case operator. It follows in particular that* $A_{N,\nu} + \nu^{2}$ *and* $\tilde{A}_{N,\nu}$

have the same spectrum.

$$\sigma(A_{N,\nu}+\nu^2) = P\sigma(A_{N,\nu}+\nu^2) = P\sigma(\tilde{A}_{N,\nu}) = \sigma(\tilde{A}_{N,\nu}) =$$

$$= \{\lambda\in\mathbb{R} \mid \lambda = (n+\nu)^2,\ n\in\mathbb{N}\cup\{0\}\},\ \text{for } \nu > -1/2.$$

Moreover, $k_\nu u_\lambda$ *is eigensolution of* $\tilde{A}_{N,\nu}$ *for eigenvalue* λ *iff* u_λ *is eigensolution of* $A_{N,\nu}+\nu^2$ *for the same eigenvalue. Similar reasoning applies to the Dirichlet case.*

2.1.3 Jacobi polynomials

Differential expression: $\mathfrak{a} := (1-m)^{-\alpha}(1+m)^{-\beta}\, 2\pi D\, (1-m)^{\alpha+1}(1+m)^{\beta+1}\, 2\pi D$.

Selfadjoint realization:

Extension $A_{N,\alpha,\beta} = C_{\alpha,\beta}\, C^*_{\alpha,\beta}$ of the operator $A_{\alpha,\beta}$ defined by $\mathfrak{a}$ applied to $C_\infty((-1,+1))$ in Hilbert space

$$H_{\alpha,\beta} = (1-m)^{-\alpha/2}\,(1+m)^{-\beta/2}\, L_2((-1,+1)),\ \alpha,\beta\in\mathbb{R}.$$

The associated operator $C_{\alpha,\beta}$ is given by

$$C_{\alpha,\beta}\colon (1-m)^{\alpha/2+1/2}(1+m)^{\beta/2+1/2}\, L_2((-1,1)) \longrightarrow H_{\alpha,\beta}, \qquad (2.37)$$

where

$$C_{\alpha,\beta}\,\varphi = \mathfrak{l}^*_{\alpha,\beta}\,\varphi \equiv -i\,(1-m)^{-\alpha}\,(1+m)^{-\beta}\,\varphi'$$

for $\varphi\in C_\infty((-1,1))$. Recall that by definition $C_\infty((-1,1))$ is dense in $D(C_{\alpha,\beta})$ with respect to the graph norm of $C_{\alpha,\beta}$. The differential expression

$$\mathfrak{l}_{\alpha,\beta} = 2\pi\,(1-m)^{\alpha+1}\,(1+m)^{\beta+1}\, D,$$

corresponds to the adjoint operator $C^*_{\alpha,\beta}$.

To reduce the number of cases to be considered we assume that

$\alpha,\ \beta > -1,\ \alpha,\ \beta \notin \mathbb{N}\cup\{0\}$.

Domain:

We have

(i) $D(A_{N,\alpha,\beta}) = \{\Phi\in H_{\alpha,\beta} \mid (1-m^2)^{1/2}\Phi' \in H_{\alpha,\beta}\,, \mathfrak{a}\Phi \in H_{\alpha,\beta}\}$, for $\alpha,\beta \notin (-1,0)$,

(ii) $D(A_{N,\alpha,\beta}) = \{\Phi\in H_{\alpha,\beta} | (1-m^2)^{1/2}\Phi' \in H_{\alpha,\beta}, a\Phi \in H_{\alpha,\beta},$

$$(1-x)^{\alpha/2+1/2}\ \Phi'(x) \longrightarrow 0 \text{ as } x \longrightarrow 1$$

$$\text{and } (1+x)^{\beta/2+1/2}\ \Phi'(x) \longrightarrow 0 \text{ as } x \longrightarrow -1\ \},$$

for $\alpha,\beta \in (-1,0)$,

(iii) $D(A_{N,\alpha,\beta}) = \{\Phi\in H_{\alpha,\beta}\ |\ (1-m^2)^{1/2}\Phi' \in H_{\alpha,\beta}, a\Phi \in H_{\alpha,\beta},$

$$(1+x)^{\beta/2+1/2}\ \Phi'(x) \longrightarrow 0 \text{ as } x \longrightarrow -1\},$$

if $\alpha \notin (-1,0)$, $\beta \in (-1,0)$,

(iv) $D(A_{N,\alpha,\beta}) = \{\Phi\in H_{\alpha,\beta}\ |\ (1-m^2)^{1/2}\Phi' \in H_{\alpha,\beta}, a\Phi \in H_{\alpha,\beta},$

$$(1-x)^{\alpha/2+1/2}\ \Phi'(x) \longrightarrow 0 \text{ as } x \longrightarrow 1\ \},$$

if $\alpha \in (-1,0)$, $\beta \notin (-1,0)$.

Since cases (iii), (iv) are unitarily equivalent with unitary equivalence featured by $\ell_{-1,0}$, we only consider (i), (ii), (iv).

Comment: The spectral analysis can be dealt with similarly as in the last case study. In fact, specializing $\alpha = \beta =: \nu - 1/2$ leads to the operator discussed in 2.1.2. We can find the necessary asymptotic results to handle this problem (following the same pattern as laid out in 2.1.1 and 2.1.2) in [23], chapter II, [25], p. 163-166. There are two linearly independent solutions of the problem

$$v_\lambda^{(\alpha,\beta)} = F(-\mu,\ \mu+\alpha+\beta+1,\ \alpha+1;\ \frac{1-x}{2})/\Gamma(\alpha+1),$$

$$w_\lambda^{(\alpha,\beta)} = (\frac{1-x}{2})^{-\alpha}\ F(-\mu-\alpha,\ \mu+\beta+1,\ 1-\alpha;\ \frac{1-x}{2})/\Gamma(1-\alpha), \tag{2.38}$$

where μ is such that

$$(\mu + \frac{\alpha+\beta+1}{2})^2 = \lambda + (\frac{\alpha+\beta+1}{2})^2,$$

and $F(a,b,c;\ z)$ is the standard principal solution of the hypergeometric equation. We may assume without loss of generality that

$$2\mu + \alpha + \beta + 1 > 0 \text{ and } \mu \neq 0 \text{ if } \lambda \in \mathbb{R}^+ \text{ and } \alpha + \beta < 0,$$

$\mu > 0$ if $\lambda \in \mathbb{R}^+$ and $\alpha + \beta \geq 0$,

$\mu = 0$ if $\lambda = 0$.

If $0 > \lambda \geq -(\frac{\alpha+\beta+1}{2})^2$ then choose the root with largest real value.

For other values of λ we choose the root μ with positive imaginary part.

Linear independence of these solutions is displayed in the fact that the Wronskian

$$W(v_\lambda^{(\alpha,\beta)}, w_\lambda^{(\alpha,\beta)}) = \frac{\sin(\pi\alpha)}{2\pi} (\frac{1+m}{2})^{-\beta-1} (\frac{1-m}{2})^{-\alpha-1},$$

is non-zero in the interval (-1,1) (for $\alpha \notin \mathbb{Z}$, the case $\alpha \in \mathbb{Z}$ can be dealt with similarly by choosing a different second solution).

The asymptotic behaviour of these solutions at +1 follows from the asymptotic properties of F at 0 (see [23], p. 37) and the equation (see [23], p. 41)

$$\frac{d}{dz} F(a,b,c;z) = \frac{a\,b}{c} F(a+1,b+1,c+1;z). \tag{2.39}$$

From the latter we have

$$\frac{d}{dx} v_\lambda^{(\alpha,\beta)}(x) = \frac{\lambda}{2} v_{\lambda+\alpha+\beta+2}^{(\alpha+1,\beta+1)}(x),$$

$$\frac{d}{dx} w_\lambda^{\alpha,\beta}(x) = (1-x)^{-1} (\alpha\, w_\lambda^{(\alpha,\beta)}(x) + (\lambda+\alpha(\beta+1))(1-\alpha)\, w_\lambda^{(\alpha-1,\beta+1)}(x)).$$

Note that the transition of α,β to $(\alpha+1)$, $(\beta+1)$ and λ to $\lambda+\alpha+\beta+2$ in the formula for $v_\lambda^{(\alpha,\beta)}$ corresponds to μ going over into $(\mu-1)$.

Altogether we get

$$\begin{aligned} &v_\lambda^{(\alpha,\beta)}(x) \sim 1\,,\quad w_\lambda^{(\alpha,\beta)}(x) \sim (1-x)^{-\alpha} \quad \text{as } x \longrightarrow 1, \\ &\frac{d}{dx} v_\lambda^{(\alpha,\beta)}(x) \sim \lambda,\quad \frac{d}{dx} w_\lambda^{(\alpha,\beta)}(x) \sim (1-x)^{-\alpha-1}, \end{aligned} \tag{2.40}$$

as $x \longrightarrow -1$, (recall the notation '$\sim$' introduced in 2.1.2). Using the linear transformation rule

$$\begin{aligned} F(a,b,c;z) = {} & \frac{\Gamma(c)\,\Gamma(c-a-b)}{\Gamma(c-a)\,\Gamma(c-b)} F(a,b,a+b-c+1;1-z) + \\ & + (1-z)^{c-a-b} \frac{\Gamma(c)\,\Gamma(a+b-c)}{\Gamma(a)\,\Gamma(b)} F(c-a,c-b,c-a-b+1;1-z), \end{aligned} \tag{2.41}$$

for $c-a-b \notin \mathbb{Z}$, $\arg(1-z) \in (-\pi,\pi)$, we obtain the asymptotic behaviour at the other boundary point ([23], p. 47). Otherwise a, b , c-a, c-b are arbitrary by analytic continuation. Thus we have for $\mu \in \mathbb{N} \cup \{0\}$ or $\mu+\alpha+\beta \in -\mathbb{N}$ or $\beta < 0$

$$v_\lambda^{(\alpha,\beta)}(x) \sim \frac{1}{\Gamma(-\mu-\beta)\ \Gamma(\mu+\alpha+1)}, \tag{2.42}$$

as $x \longrightarrow -1$. For $\beta > 0$ and $\mu \notin \mathbb{N} \cup \{0\}$ and $\mu+\alpha+\beta \notin -\mathbb{N}$

$$v_\lambda^{(\alpha,\beta)}(x) \sim (1+x)^{-\beta} \text{ as } x \longrightarrow -1. \tag{2.43}$$

For $w_\lambda^{(\alpha,\beta)}$ we obtain

$$w_\lambda^{(\alpha,\beta)}(x) \sim \frac{1}{\Gamma(-\mu-\alpha-\beta)}, \tag{2.44}$$

if $\beta < 0$ or $\mu+\beta \in -\mathbb{N}$ or $\mu+\alpha \in \mathbb{N} \cup \{0\}$ as $x \longrightarrow -1$; and

$$w_\lambda^{(\alpha,\beta)}(x) \sim (1+x)^{-\beta} \quad \text{as } x \longrightarrow -1, \tag{2.45}$$

if and $\beta > 0$ and $\mu+\beta \notin -\mathbb{N}$ and $\mu+\alpha \notin \mathbb{N} \cup \{0\}$. Also,

$$\frac{d}{dx} v_\lambda^{(\alpha,\beta)}(x) \sim \lambda \quad \text{as } x \longrightarrow -1, \tag{2.46}$$

if $\mu \in \mathbb{N} \cup \{0\}$ or $\mu + \alpha + \beta + 1 \in -\mathbb{N}$; for $\mu \notin \mathbb{N} \cup \{0\}$ and $\mu+\beta \notin \mathbb{N} \cup \{0\}$

$$\frac{d}{dx} v_\lambda^{(\alpha,\beta)}(x) \sim (1+x)^{-\beta-1}, \tag{2.47}$$

as $x \longrightarrow -1$; for $\mu+\alpha \notin \mathbb{N} \cup \{0\}$ and $\mu + \alpha + \beta + 1 \notin -\mathbb{N}$

$$\frac{d}{dx} w_\lambda^{(\alpha,\beta)}(x) \sim (1+x)^{-\beta-1}, \tag{2.48}$$

as $x \longrightarrow -1$. From these asymptotic statements we can derive as before that for $\alpha,\ \beta > 0$

$$B = C^*. \tag{2.49}$$

This implies the characterization of $D(A_{N,\alpha,\beta})$ given for this case. Similar to (2.34) we have the following Sobolev type estimate

$|\varphi(t)-\varphi(s)|^2 \leq$

$$\leq \int_{(s,t)} (1-x)^{\alpha} (1+x)^{\beta} dx \int_{(s,t)} (1-x)^{-\alpha}(1+x)^{-\beta} |\varphi'(x)|^2 dt, \qquad (2.50)$$

for $\varphi \in D(B^*_{\alpha,\beta})$, $s \leq t$. This implies continuity of any function u in $D(B^*_{\alpha,\beta})$ near the boundary points ± 1 and the boundary condition

$$\lim_{x \to \pm 1 \mp} (1-x)^{-\alpha/2-1/2} (1+x)^{-\beta/2-1/2} u(x) = 0, \qquad (2.51)$$

for $u \in D(C_{\alpha,\beta})$. This yields the boundary condition

$$\lim_{x \to \pm 1 \mp} (1-x)^{-\alpha/2-1/2} (1+x)^{-\beta/2-1/2} (\ell\, u)(x) = 0,$$

for $u \in D(A_{N,\alpha,\beta})$. Since

$$\ell\, \varphi = (1-m)^{\alpha+1} (1+m)^{\beta+1}\, 2\pi D\, \varphi,$$

for $\varphi \in D(C^*_{\alpha,\beta})$, the boundary condition becomes

$$\lim_{x \to \pm 1 \mp} (1-x)^{\alpha/2+1/2} (1+x)^{\beta/2+1/2} u'(x) = 0, \qquad (2.52)$$

for $u \in D(A_{N,\alpha,\beta})$. Applying the same reasoning as in section 2.1.2. we see that an element u in the ortho-complement of $D(C_{\alpha,\beta})$ as a subspace of

$$\{\Phi \in D(B^*_{\alpha,\beta}) \mid \lim_{x \to \pm 1 \mp} (1-x)^{-\alpha/2-1/2} (1+x)^{-\beta/2-1/2} \Phi(x) = 0\}, \qquad (2.53)$$

with respect to the graph inner product of $B^*_{\alpha,\beta}$ has to satisfy

$$(\ell_{\alpha,\beta}\, \ell^*_{\alpha,\beta} + 1)\, u = 0.$$

Since neither of the possible candidates

$$\ell_{\alpha,\beta}\, v_\lambda^{(\alpha,\beta)} = -\, i\lambda/2\, (1-m)^{\alpha+1} (1+m)^{\beta+1}\, v^{(\alpha+1,\beta+1)}_{\lambda+\alpha+\beta+2},$$

$$\ell_{\alpha,\beta}\, w_\lambda^{(\alpha,\beta)} = -i\, (1-m)^{\alpha} (1+m)^{\beta+1} (\alpha\, w_\lambda^{(\alpha,\beta)} + \lambda\, (1-\alpha)\, w_\lambda^{(\alpha-1,\beta+1)}),$$

does satisfy both the boundary condition and the condition to be an element of $D(B^*_{\alpha,\beta})$, we have

$$D(C_{\alpha,\beta}) = \tag{2.54}$$

$$= \{\Phi \in D(B^*_{\alpha,\beta}) \mid \lim_{x \to \pm 1 \mp} (1-x)^{-\alpha/2-1/2} (1+x)^{-\beta/2-1/2} \Phi(x) = 0\}.$$

This implies the above description of $D(A_{N,\alpha,\beta})$ in the case $\alpha, \beta \in (-1,0)$.

Equation (2.49) indicates that the boundary condition at +1 (or -1) is only needed when $\alpha \in (-1,0)$, (or $\beta \in (-1,0)$). The details are left as an exercise.

Spectrum: $\sigma(A_{N,\alpha,\beta}) = P\sigma(A_{N,\alpha,\beta}) = \{\lambda \in \mathbb{R} \mid \lambda = n(n+\alpha+\beta+1) \ , \ n \in \mathbb{N} \cup \{0\}\}$.

Comment: That for $\alpha, \beta > 0$

$$P\sigma(A_{N,\alpha,\beta}) = \{\lambda \in \mathbb{R} \mid \lambda = n(n+\alpha+\beta+1) \ , \ n \in \mathbb{N} \cup \{0\}\}$$

is again clear from the asymptotic behaviour of candidates for eigensolutions at the boundary. To show that the rest of the real axis is resolvent set follows along the lines of previous considerations of the same kind. In the other cases we have to take the boundary condition at boundary point +1 or -1 into account.

Eigensolutions:

$$N(A_{N,\alpha,\beta}-\lambda) = \mathrm{span}\{v_\lambda^{(\alpha,\beta)}\} \ , \ \lambda = n(n+\alpha+\beta+1) \ , \ n \in \mathbb{N} \cup \{0\}, \ \alpha, \beta > -1.$$

The functions $v_n^{(\alpha,\beta)}$ are polynomials. The polynomials

$$P_\lambda^{(\alpha,\beta)} = (-1)^\lambda \begin{bmatrix} \lambda+\beta \\ \lambda \end{bmatrix} v_\lambda^{(\alpha,\beta)} \ , \ \lambda \in \mathbb{N} \cup \{0\},$$

are known as Jacobi polynomials.

The set

$$\left\{ \left[\frac{n!\Gamma(\alpha+\beta+n+1)(\alpha+\beta+2n+1)}{2^{\alpha+\beta+1}\Gamma(\alpha+n+1)\Gamma(\beta+n+1)} \right]^{1/2} P_n^{(\alpha,\beta)} \right\}_{n\in\mathbb{N}\cup\{0\}},$$

is a complete, orthonormal set of eigensolutions in $H_{\alpha,\beta}$.

<u>Remark 2.2</u>: *The case* $\alpha = \beta = \nu - 1/2$ *leads to the situation discussed in 2.1.2,* $\alpha = \beta = 0$ *leads to Legendre polynomials as considered in 2.1.1*

and the case $\alpha = \beta = -1/2$ leads to Chebycheff polynomials. The spectral results for the latter case will be stated separately in the next section.

2.1.4 Chebycheff polynomials

Differential expression: $\mathfrak{a} := (1-m^2)^{1/2}\, 2\pi D\, (1-m^2)^{1/2}\, 2\pi D$.

Selfadjoint realization:

Special case $A = A_{N,-1/2,-1/2}$ of section 2.1.3. Underlying Hilbert space is $H = H_{-1/2,-1/2}$ in the notation of 2.1.3.

Domain:

$$D(A) = \{\Phi \in H \mid (1-m^2)^{1/2}\,\Phi' \in H,\ \mathfrak{a}\Phi \in H, \text{ and } (1-x^2)^{1/4}\Phi'(x) \longrightarrow 0 \text{ as } x \longrightarrow \pm 1\}.$$

Spectrum: $\sigma(A) = P\sigma(A) = \{t \in \mathbb{R} \mid t = n^2,\ n \in \mathbb{N} \cup \{0\}\}$.

Eigensolutions:

$N(A-\lambda) = \mathrm{span}\langle T_n\rangle$, $\lambda = n^2$, $n \in \mathbb{N} \cup \{0\}$, where T_n denotes the Chebycheff polynomial of order n . If E_n denotes the n-th Euler number, then $\left\{ \left[\frac{E_n}{\pi}\right]^{1/2} T_n \right\}_{n \in \mathbb{N}\cup\{0\}}$ is a complete, orthonormal set in H . It is

$$T_n = 2^{2n}\,\frac{(n!)^2}{(2n)!}\,P_n^{(-1/2,-1/2)} , \quad n \in \mathbb{N} \cup \{0\},$$

(see Remark 2.2).

<u>Remark 2.3</u>: *Since we have $\alpha = \beta$ in this case, Chebycheff polynomials may also be considered as special Gegenbauer polynomials. We have*

$$T_n = \frac{n}{2}\, C_n^0 , \quad n \in \mathbb{N} \cup \{0\}.$$

2.1.5 Hermite polynomials

Differential operator: $A := \mathfrak{D}\,\mathfrak{D}^*$ (i.e. $C \equiv \mathfrak{D}$) with

$$D(A) = \mathcal{S}_2 \subset L_2(\mathbb{R}). \tag{2.55}$$

Operator A is selfadjoint by definition. It coincides with the closure A of $A_{\restriction \mathring{C}_\infty(\mathbb{R})}$.

Spectrum: $\sigma(A) = P\sigma(A) = \mathbb{N}/\pi \cup \{0\}$.

Comment: To show that the point spectrum is equal to $\mathbb{N}/\pi \cup \{0\}$ and that there is no other spectrum, we shall employ as a short-cut a different approach. If we can give a complete system of eigensolutions — say an orthonormal system of eigensolutions — then there cannot be other spectral contributions. This follows directly from the spectral theorem. This reasoning could have been used in the other cases of classical polynomials as well. A presentation that is more in keeping with the ideas previously used can be found in [36], paragraph 24. According to Remark 1.8 the functions $\mathscr{h}_k$, $k \in \mathbb{N} \cup \{0\}$, form a complete orthonormal set. Moreover, according to (1.69) and Remark 1.8 we have

$$\mathscr{D}\,\mathscr{D}^* \mathscr{h}_k = \frac{k}{\pi} \mathscr{h}_k. \tag{2.56}$$

This proves the above statement about the spectrum of A. Note in particular that again by (1.68)

$$\sigma(\mathscr{D}\,\mathscr{D}^*) + \frac{1}{\pi} = \sigma(\mathscr{D}^*\mathscr{D}) = \mathbb{N}/\pi. \tag{2.57}$$

Eigensolutions:

$N(A-k) = \operatorname{span}\{\mathscr{h}_k\}$, $k \in \mathbb{N} \cup \{0\}$. The polynomial H_k defined by

$$H_k(x) = (-i)^k\, 2^{k/2-1/4}\, (k!)^{1/2}\, \gamma^{-1}(x/(2\pi)^{1/2})\, \mathscr{h}_k(x/(2\pi)^{1/2}),\ x \in \mathbb{R},$$

is known as Hermite polynomial of order k (see Remark 1.8), where $k \in \mathbb{N} \cup \{0\}$.

2.1.6 Laguerre polynomials

Differential expression: $\mathscr{a} := m^{-\alpha}\, e^{m}\, 2\pi D\, m^{\alpha+1}\, e^{-m}\, 2\pi D$, $\alpha > -1$ fixed.

Selfadjoint realization:

Operator $A = A_{N,\alpha} = C_\alpha C_\alpha^*$ where

$$C_\alpha\colon m^{\alpha/2}\, e^{-m/2}\, L_2(\mathbb{R}^+) \longrightarrow m^{-\alpha/2}\, e^{m/2}\, L_2(\mathbb{R}^+), \tag{2.58}$$

with

$$C_\alpha \varphi = \mathscr{c}_\alpha^* \varphi = -i\, m^{-\alpha}\, e^{m}\, \varphi', \tag{2.59}$$

for $\varphi \in D(C_\alpha)$. The domain $D(C_\alpha)$ is — as usual — the completion of $C_\infty(\mathbb{R}^+)$ with respect to the graph norm of C_α. The underlying Hilbert space H for the selfadjoint operator A is

$$H = m^{-\alpha/2}\, e^{m/2}\, L_2(\mathbb{R}^+).$$

The formal differential operator

$$\mathfrak{c} = 2\pi\, m^{\alpha+1}\, e^m\, D,$$

corresponds to the adjoint operator C_α^*.

The differential equation

$$(\mathfrak{a} - \lambda)\, u = 0,$$

is known as Kummer's differential equation or confluent hypergeometric equation (see [23], p. 268 pp). Let us assume for simplicity that $\alpha > -1$.

Domain: $D(A) = \{ \Phi \in H \mid m^{1/2}\, \Phi' \in H\ ,\ \mathfrak{a}\, \Phi \in H\ ,$ and

$$\lim_{x \to 0+} x^{(\alpha+1)/2}\, \Phi'(x) = 0\}\ \text{for}\ \alpha \in (-1,0).$$

Comment: For $\lambda \notin \mathbb{N} \cup \{0\}$ two linearly independent solutions are provided by

$$W_{\lambda,1} = \Phi(-\lambda,\alpha+1;x)/\Gamma(\alpha+1),$$

$$W_{\lambda,2}(x) = \frac{\pi}{\sin(\pi(\alpha+1))} \left\{ \frac{W_{\lambda,1}(x)}{\Gamma(-\lambda-\alpha)} - x^{-\alpha} \frac{\Phi(-\lambda-\alpha,1-\alpha;x)}{\Gamma(1-\alpha)\ \Gamma(-\lambda)} \right\}.$$

where $\Phi(a,c;z)$ denotes Kummer's function. All these are well-defined (by analytic continuation) for any values of λ,α. For their Wronskian we have

$$W(W_{\lambda,2}, W_{\lambda,1}) = e^m\, m^{-\alpha-1}/\Gamma(-\lambda). \qquad (2.60)$$

As asymptotic behaviour we find

$$\left.\begin{array}{l}
W_{\lambda,1}(x) \sim e^x\, x^{-(\lambda+\alpha+1)} \quad \text{as } x \longrightarrow \infty\ ,\ \lambda \notin \mathbb{N} \cup \{0\}, \\
W_{\lambda,2}(x) \sim x^{\lambda} \text{ as } x \longrightarrow \infty, \\
W_{\lambda,1}(x) \sim 1 \text{ as } x \longrightarrow 0, \\
\text{and} \\
W_{\lambda,2}(x) \sim \dfrac{1}{\Gamma(-\lambda-\alpha)} \text{ as } x \longrightarrow 0, \\
\text{for } \alpha \in (-1,0),
\end{array}\right\} \qquad (2.61)$$

$$\left.\begin{aligned} & W_{\lambda,2}(x) \sim x^{-\alpha} \text{ as } x \longrightarrow 0,\ \lambda \notin \mathbb{N} \cup \{0\}, \\ & \text{for } \alpha > 0; \\ & W_{\lambda,2}(x) \sim -\frac{\ln x}{\Gamma(-\lambda)} \text{ as } x \longrightarrow 0, \\ & \text{for } \alpha = 0. \end{aligned}\right\} \quad (2.61)$$

For $\lambda \in \mathbb{N} \cup \{0\}$ the function $W_{\lambda,2}$ (also $W_{\lambda,1}$) is a polynomial. Corresponding asymptotic results for the derivatives can be obtained from

$$\Phi'(a,c;x) = a/c\ \Phi(a+1,c+1;x), \quad (2.62)$$

(see [23], p. 264). We have for $\alpha > -1$ a Sobolev type estimate

$$|\varphi(s)-\varphi(t)|^2 \leq \frac{|s^{\alpha+1} - t^{\alpha+1}|}{\alpha + 1} \int_{(s,t)} x^{-\alpha} e^x |\Phi'(x)|^2 \, dx, \quad (2.63)$$

for $\varphi \in D(B_\alpha^*)$, $t \geq s \geq 0$. Thus if Φ is in the domain of the associated operator C_α, then it has to satisfy the boundary condition

$$\lim_{x \longrightarrow 0+} x^{-(\alpha+1)/2} \Phi(x) = 0.$$

For elements $u \in D(A)$ this implies the Neumann type boundary condition

$$\lim_{x \longrightarrow 0+} x^{(\alpha+1)/2} u'(x) = 0, \quad (2.64)$$

if $\alpha \in (-1,0)$. By the techniques used in previous cases we can show that

$$D(C_\alpha) = \{\Phi \in D(B_\alpha^*) \mid \lim_{x \longrightarrow 0+} x^{-(\alpha+1)/2} \Phi(x) = 0\},$$

which implies the representation of $D(A)$.

Spectrum: $\sigma(A) = P\sigma(A) = \mathbb{N} \cup \{0\}$.

Comment: he Polynomials $W_{\lambda,2}$ (or $W_{\lambda,1}$), $\lambda \in \mathbb{N} \cup \{0\}$, are apparently eigen-solutions. For $\lambda \in \mathbb{N} \cup \{0\}$ we need a second linearly independent solution which can be taken to be

$$W_{\lambda,3}(x) = x^{-\alpha}\, U(-\lambda-\alpha, 1-\alpha; x), \ x \in \mathbb{R}^+,$$

which does, however, not satisfy the boundary condition. Therefore the spectrum is simple. All other points of the real axis are in the resolvent set as follows from investigating the integral representation of the resolvent in terms of Green's function. Green's function

$$G_\lambda(x,y) = c\,(h(x-y)\,u_\lambda(x)\,v_\lambda(y) + h(y-x)\,u_\lambda(y)\,v_\lambda(x)),$$

is given by $u_\lambda = W_{\lambda,2}$ and $v_\lambda = W_{\lambda,1}$ and

$$c = \Gamma(-\lambda).$$

Eigensolutions:

$N(A-\lambda) = \text{span}\langle W_{\lambda,2}\rangle$, $\lambda \in \mathbb{N} \cup \{0\}$. The polynomials

$$L_m^{(\alpha)} \equiv \frac{(-1)^\lambda}{\lambda!} W_{\lambda,2}, \quad \lambda \in \mathbb{N} \cup \{0\},$$

are the Laguerre polynomials of index $\alpha > -1$ and order m .

The set

$$\left\{ \frac{1}{\Gamma(1+\alpha)\begin{bmatrix}\lambda+\alpha\\ \lambda\end{bmatrix}} L_\lambda^{(\alpha)} \right\}_{\lambda\in\mathbb{N}\cup\{0\}},$$

is a complete, orthonormal family of eigensolutions of A in H.

2.2 Spectral results related to the differential expression D^2. The one-sided Fourier sine and Fourier cosine transform

In 1.1.3.1 we have already considered the differential expression D^2 defined on $C_\infty(\mathbb{R})$. We shall reconsider D^2, now, from a different point of view. We recall from 1.1.3.1 that

$$f \longrightarrow j_s^{-1}(\mathcal{F}_{cos}(f)(\sqrt{t})t^{-1/4}(\lambda) \tag{2.65}$$

is a unitary mapping from $\operatorname{sym}(L_2(\mathbb{R}))$ onto $L_2(\mathbb{R}^+)$; in fact, it is a spectral representation for $A = D^2$ restricted to $\operatorname{sym}(L_2(\mathbb{R}))$. Here D denotes the selfadjoint operator of section 1.1.2. Since j_s^{-1} is a unitary mapping from $\operatorname{sym}(L_2(\mathbb{R}))$ onto $L_2(\mathbb{R}^+)$, we immediately see that

$$F_{cos}: L_2(\mathbb{R}^+) \longrightarrow L_2(\mathbb{R}^+)$$

$$f \longrightarrow j_s^{-1} \mathcal{F}_{cos}(j_s(f)(\sqrt{t})t^{-1/4} \tag{2.66}$$

is unitary. We obtain for F_{cos} in integral form

$$F_{cos}(f)(\lambda) = (\tfrac{2}{\pi})^{1/2} \lambda^{-1/4} \int_{\mathbb{R}^+} \cos(\sqrt{\lambda}\bullet x)\ f(x)\ dx \ . \tag{2.67}$$

We shall see that F_{cos} is a spectral representation for $A = C\,C^*$ where

$$C : D(C) \subset L_2(\mathbb{R}^+) \longrightarrow L_2(\mathbb{R}^+) ,$$

$$u \longrightarrow D^2 u ,$$

with

$$D(C) := \{\Phi \in H_1(\mathbb{R}^+) \mid \Phi_{\restriction 0} = 0\} \equiv H_1(\mathbb{R}^+).$$

Note that by definition we have that $C_\infty(\mathbb{R}^+)$ is dense in domain of C. First, we show that

$$j_s D(C) \subset D(D). \tag{2.68}$$

If $\Phi \in C_\infty(\mathbb{R}^+) \subset D(C)$ then

$$\Phi(x) = - \int_x^\infty \Phi'(t)\, dt.$$

By definition it follows that

$$j_s\Phi(x) = \begin{cases} - \int_x^\infty \Phi'(t)\, dt & \text{if } x \geq 0 \\ - \int_{-x}^\infty \Phi'(t)\, dt & \text{if } x \leq 0 \end{cases},$$

and thus

$$j_s\Phi \in C_\infty(\mathbb{R}).$$

In particular

$$j_s\Phi \in \text{sym}(C_\infty(\mathbb{R})). \tag{2.69}$$

Relation (2.69) proves (2.68) by taking closures.

We have according to (1.146)

$$F_{\cos}(A\Phi)(\lambda) = \lambda\, F_{\cos}(\Phi)(\lambda),\ \lambda \in \overline{\mathbb{R}^+}, \tag{2.70}$$

for all $\Phi \in D(A)$. Since the selfadjointness of A follows from the selfadjointness of $D^2{\restriction}_{\text{sym}(L_2(\mathbb{R}))}$ and (2.69) we have proved

<u>Lemma 2.1</u>: $F_{\cos}$ *is a spectral representation for* A .

By Lemma 2.1 the spectral results for A can be obtained from those of 1.1.3.1.

<u>Theorem 2.1</u>: A *is selfadjoint and has a simple spectrum:*

$$\sigma(A) = C\ \sigma(A) = \overline{\mathbb{R}^+} .$$

The unitary mapping $F_{\cos}$ *is a spectral representation for* A.

<u>Remark 2.4</u>: *The domain of* A *can be characterized as*

$$D(A) = \{\Phi \in H_1(\mathbb{R}^+) \mid \Phi' \in H_1(\mathbb{R}^+)\}.$$

Similar arguments could be used to develop a spectral representation for

$$\tilde{A} : D(\tilde{A}) \subset L_2(\mathbb{R}^+) \longrightarrow L_2(\mathbb{R}^+)$$
$$u \longrightarrow D^2 u ,$$

with

$$D(\tilde{A}) = \{\Phi \in H_1(\mathbb{R}^+) \mid \Phi' \in H_1(\mathbb{R}^+)\},$$

by considering the antisymmetric part. We have

$$\tilde{A} = C^* C.$$

We state the corresponding spectral result as

<u>Theorem 2.2</u>: $\tilde{A}$ *is selfadjoint and has simple spectrum*

$$\sigma(\tilde{A}) = C\,\sigma(\tilde{A}) = \mathbb{R}^+.$$

The mapping $F_{sin}(f)(\lambda) := j_a^{-1}\, \mathcal{F}_{sin}(j_a(f))(\sqrt{\lambda})\, \lambda^{-1/4}$, $f \in L_2(\mathbb{R})$, *is a spectral representation of* $\tilde{A}$.

<u>Remark 2.5</u>: F_{cos} , F_{sin} *are the <u>one-sided</u> <u>Fourier-cosine</u> <u>and</u> <u>Fourier-sine</u> <u>transform</u> respectively.*

We will not consider the discrete case here, since it is an easy exercise. We note however that considering D^2 with the Dirichlet boundary condition on the boundary of the underlying domain I (say $I = (-1/2,1/2)$) would lead to a pure sine-series, where the Neumann case would yield a pure cosine series expansion as spectral theorems.

As pointed out before, in the following we shall restrict our attention to the case of continuous spectrum only. Another restriction is, that we consider merely the Neumann case in the sense of the general remarks at the beginning of part 2. In as much as the reconstruction of integral transforms is concerned other boundary conditions seem to be of a lesser interest. There is however no problem in dealing with other boundary conditions in a similar fashion.

2.3 The Hankel transform as a spectral representation related to the differential expression $m^{-1}\, 2\pi D\, m\, 2\pi D + \nu^2 m^{-2}$, $\nu \in \mathbb{R}$

Our second example appears in connection with separation of variables of the two-dimensional Laplacian in polar coordinates. The differential expression $\mathcal{a}_\nu := m^{-1}\, 2\pi D\, m\, 2\pi D + \nu^2 m^{-2}$ ($\nu \geq 0$ without loss of generality) does not seem to be of the general form considered so far. Letting

$$\mathcal{b} = m^{\nu+1} (2\pi D)\, m^{-\nu},$$

shows, however, by an elementary calculation that

$$\mathcal{a}_\nu = \mathcal{b}^* \mathcal{b},$$

where the formal adjoint $\mathcal{b}^*$ is taken with respect to the inner products of $H = m^{-1/2} L_2(\mathbb{R}^+)$ as initial space and $m^{1/2} L_2(\mathbb{R}^+)$ as terminal space for $\mathcal{b}$. Let us consider the Neumann case. In the notation of (2.5) it is

$$q = m, \ \rho = m^\nu.$$

The operator $A_{N,\nu} = C_\nu C_\nu^*$ where

$$C_\nu : D(C_\nu) \subset m^{1/2} L_2(\mathbb{R}^+) \longrightarrow m^{-1/2} L_2(\mathbb{R}^+), \tag{2.71}$$

with

$$C_\nu \varphi = b^* \varphi = m^{-\nu-1}\, 2\pi D\, m^\nu \varphi, \ \varphi \in C_\infty(\mathbb{R}^+). \tag{2.72}$$

By the generic definition (2.7) of C_ν we know that $C_\infty(\mathbb{R}^+)$ is dense in the domain of C_ν. As before a Sobolev type estimate shows us the boundary condition satisfied by elements in $D(C_\nu)$. We have for an element $\Phi \in D(B_\nu^*)$ (compare generic definition (2.6))

$$|s^\nu \Phi(s) - t^\nu \Phi(t)|^2 \leq \frac{|s^{2(\nu+1)} - t^{2(\nu+1)}|}{2(\nu+1)} \int_{(s,t)} x\, |(b^*\Phi)(x)|^2\, dx, \tag{2.73}$$

for $t \geq s > 0$. This implies the boundary condition

$$x^{-1}\, \Phi(x) \longrightarrow 0 \text{ as } x \longrightarrow 0+,$$

for elements in $D(C_\nu)$. As a consequence we have

$$x^{\nu}\,(x^{-\nu}\,u(x))' \longrightarrow 0 \quad \text{as } x \longrightarrow 0+, \tag{2.74}$$

for $u \in D(A_{N,\nu})$ (compare [36], p. 353). If we consider the ortho-complement $D(C_\nu)^{\perp}$ of $D(C_\nu)$ in

$$\{\Phi \in D(B_\nu^{*}) \mid \lim_{x \to 0+} x^{-1}\,\Phi(x) = 0\}, \tag{2.75}$$

with respect to the graph inner product of B_ν^{*}, we see that an element v in $D(C_\nu)^{\perp}$ must satisfy

$$(\ell\,\ell^{*} + 1)\,v = 0. \tag{2.76}$$

Candidates for solutions are $\ell\, I_\nu = -i\,m\,I_{\nu+1}$, $\ell\, K_\nu = i\,m\,K_{\nu+1}$, where I_ν, K_ν are the modified Bessel functions (see e.g. [23], p. 66). It is

$$I_\nu(x) = J_\nu(ix), \quad K_\nu(x) = \frac{1}{2}\,i\pi\,e^{i\pi\nu/2}\,H_\nu^{+}(ix), \quad x \in \mathbb{R}, \tag{2.77}$$

where J_ν, H_ν^{+} are the Bessel function and the Hankel function of first kind, respectively. The Hankel function of first kind is related to the Neumann function N_ν by

$$H_\nu^{+}(z) = J_\nu(z) + i\,N_\nu(z), \quad z \in \mathbb{C}. \tag{2.78}$$

Also note that the Hankel function of second kind H_ν^{-} can be given as

$$H_\nu^{-}(z) = J_\nu(z) - i\,N_\nu(z), \quad z \in \mathbb{C}. \tag{2.79}$$

In the evaluation of $\ell\, I_\nu$, $\ell\, K_\nu$ we have made use of the formulas

$$\begin{aligned} 2\pi i\, m^{-1}\, D\,(m^{-\nu}\, I_\nu) &= m^{-\nu-1}\, I_{\nu+1}, \\ 2\pi i\, m^{-1}\, D\,(m^{-\nu}\, K_\nu) &= -\, m^{-\nu-1}\, K_{\nu+1}, \end{aligned} \tag{2.80}$$

(see [23], p. 67). Since we have

$$\begin{aligned} I_\nu(x) &\sim x^{\nu}, \quad \nu \in \mathbb{C}, \\ K_\nu(x) &\sim x^{-\nu} - x^{\nu} \text{ if } \nu \in \mathbb{C},\ \nu \notin \mathbb{Z}, \\ K_\nu(x) &\sim \ln(x)\, x^{-|\nu|} \text{ if } \nu \in \mathbb{Z}, \end{aligned} \tag{2.81}$$

as $x \longrightarrow 0+$, the solution $\ell\, I_\nu$ (note $\nu \geq 0$) is the only solution of (2.76) satisfying the boundary condition at zero. But because of the asymptotic behaviour at ∞ (see [23], p. 139)

$$I_\nu(x) \sim x^{-1/2} e^x \text{ as } x \longrightarrow \infty, \ \nu \in \mathbb{C}, \tag{2.82}$$

it is not in $m^{1/2} L_2(\mathbb{R}^+)$ and a-fortiori not in $D(B_\nu^*)$. Thus, we have shown that

$$D(C_\nu) = \{\Phi \in D(B_\nu^*) \mid x^{-1} \Phi(x) \longrightarrow 0 \text{ as } x \longrightarrow 0+\}.$$

As a consequence we have as a description of $D(A_{N,\nu})$

$$D(A_{N,\nu}) = \{\Phi \in H \mid \mathcal{b}\, \Phi \in m^{1/2} L_2(\mathbb{R}^+),\ \mathcal{a}_\nu\, \Phi \in H \text{ and } \lim_{x \longrightarrow 0+} x^\nu (x^{-\nu} u(x))' = 0 \}.$$

(here — as usual — application of $\mathcal{a}$, $\mathcal{b}$ has to be understood in the weak sense, i.e. in the sense of the definition of A^*, C^*).

To proceed we note the following asymptotic results:

$$\begin{aligned} &N_0(z) \sim \ln z,\ N_0'(z) \sim z^{-1}, \\ &J_0(z) \sim 1, \\ &J_\nu(z) \sim z^\nu, \quad \nu \neq 0, \\ &N_\nu(z) \sim z^{-\nu}, \quad \nu \neq 0, \quad \text{as } z \longrightarrow 0, \end{aligned} \tag{2.83}$$

and

$$H_\nu^\pm(z) := J_\nu(z) \pm i\, N_\nu(z) \sim e^{\pm iz} z^{-1/2} \quad \text{as } |z| \longrightarrow \infty, \tag{2.84}$$

for $z \in \mathbb{C}$, $\arg(z) \in (-\pi,\pi)$ (see [23], p. 139). Thus, the deficiency indices of A_μ are $(0|0)$ for $\nu \in [0,\infty)$. This finally demonstrates the validity of the description

$$D(A_{N,\nu}) = \{\Phi \in H \mid \mathcal{b}\, \Phi \in m^{1/2} L_2(\mathbb{R}^+),\ \mathcal{a}_\nu\, \Phi \in H \}.$$

for $D(A_{N,\nu})$ and that A_ν is already selfadjoint.

As in chapter 1.1.2 we shall now calculate the spectral family of $A_{N,\nu}$ via Stone's formula. Green's function $G_{\nu,\lambda}(x,t)$ will be used to represent the resolvent of $A \equiv A_{N,\nu}$:

$$((A-(\lambda \pm i\varepsilon))^{-1} f)(x) = \int_{\mathbb{R}^+} G_{\nu,\lambda \pm i\varepsilon}(x,t)\, f(t)\, dt, \quad \lambda \in \mathbb{R}, \quad \varepsilon > 0,$$

where according to (2.75), (2.76)

$$G_{\nu,\mu}(x,t) := \frac{i\pi}{2} \begin{cases} J_\nu(\sqrt{\mu}\, x)\ H_\nu^+(\sqrt{\mu}\, t) & \text{if } 0 \le x \le t \\ J_\nu(\sqrt{\mu}\, t)\ H_\nu^+(\sqrt{\mu}\, x) & \text{if } 0 \le t \le x \end{cases},$$

for $\mu \in \mathbb{C}\backslash\mathbb{R}$, (here as always $\sqrt{\mu}$ is chosen such that $\operatorname{Im}\sqrt{\mu} > 0$), $f \in H$, $\operatorname{supp} f \subset\subset \mathbb{R}$ ('$A \subset\subset B$' is an abreviation for '$\bar{A} \subset B$, $\bar{A}$ compact').

For $\lambda \in \mathbb{R}^+$, $\varepsilon > 0$, we have

$$G_{\nu,\lambda+i\varepsilon}(x,t) - G_{\nu,\lambda-i\varepsilon}(x,t) =$$

$$= \frac{i\pi}{2} \left\{ \begin{array}{ll} J_\nu(\sqrt{\lambda}\, x)\ H_\nu^+(\sqrt{\lambda}\, t) - J_\nu(-\sqrt{\lambda}\, x)\ H_\nu^+(-\sqrt{\lambda}\, t) & \text{if } 0 \le x \le t \\ J_\nu(\sqrt{\lambda}\, t)\ H_\nu^+(\sqrt{\lambda}\, x) + J_\nu(-\sqrt{\lambda}\, t)\ H_\nu^+(-\sqrt{\lambda}\, x) & \text{if } 0 \le t \le x \end{array} \right\} + \tag{2.85}$$

$$+\ o(1), \text{ as } \varepsilon \longrightarrow 0+.$$

Using

$$\begin{aligned} J_\nu(-z) &= e^{i\nu\pi}\ J_\nu(z), \\ H_\nu^+(-z) &= -\ e^{-i\nu\pi}\ H_\nu^-(z), \\ 2\ J_\nu(z) &= H_\nu^+(z) + H_\nu^-(z), \quad \nu \in \mathbb{R}, \end{aligned} \tag{2.86}$$

we conclude from (2.85)

$$G_{\nu,\lambda+i\varepsilon}(x,t) - G_{\nu,\lambda-i\varepsilon}(x,t) = i\pi\ J_\nu(\sqrt{\lambda}\, x)\ J_\nu(\sqrt{\lambda}\, t) + o(1), \tag{2.87}$$

as $\varepsilon \longrightarrow 0+$. By similar considerations as in 1.1.2 we obtain $(0 < \mu \le \lambda)$

$$\|(\pi(\lambda) - \pi(\mu))f\|_H^2 = \int_\mu^\lambda |H_\nu(f)(\sigma)|^2\, d\sigma, \tag{2.88}$$

with

$$H_\nu(f)(\sigma) = \frac{1}{\sqrt{2}} \int_{\mathbb{R}^+} t\ J_\nu(\sqrt{\sigma}\, t)\ f(t)\ dt, \tag{2.89}$$

$f \in H$, $\operatorname{supp} f \subset\subset \mathbb{R}^+$.

H_ν can be extended to all $f \in H$ by (2.88) ($\lambda \to +\infty$, $\mu \to -\infty$, note that $\mathbb{R}^-$ is not in the spectrum and that 0 is not an eigenvalue):

$$\|f\|_H^2 = \|H_\nu(f)\|^2_{L_2(\mathbb{R}^+)} := \int_{\mathbb{R}^+} |H_\nu(f)(\lambda)|^2 \, d\lambda \ , \quad f \in H. \tag{2.90}$$

The mapping

$$H_\nu : H \longrightarrow L_2(\mathbb{R}^+) \tag{2.91}$$

is isometric.

With the unitary mapping

$$k : L_2(\mathbb{R}^+) \longrightarrow H$$
$$u(t) \longrightarrow \sqrt{2}\, u(t^2) \ ,$$

an easy calculation shows that

$$H_\nu^* = k\, H_\nu\, k \ . \tag{2.92}$$

Relation (2.92) proves that H_ν is onto.

<u>Theorem 2.3</u>: *The spectrum of* $A \equiv A_{N,\nu}$ *is simple:*

$$\sigma(A) = C\,\sigma(A) = \overline{\mathbb{R}^+} \ .$$

H_ν *is a spectral representation for* A.

<u>Proof</u>: The result follows since from (2.90) and by polarization of (2.88) we have

$$H_\nu(A_\nu \Phi)(\lambda) = \lambda\, H_\nu(\Phi)(\lambda) \ , \quad \Phi \in D(A) \ .$$

□

From (2.92) we see that

$$Ha_\nu := k\, H_\nu \tag{2.93}$$

is a unitary transformation of $H \longrightarrow H$ called <u>Hankel transform</u>.

Ha_ν may be written in integral form as

$$Ha_\nu(f)(\lambda) = \int_{\mathbb{R}^+} t\, J_\nu(\lambda\cdot t) f(t)\, dt \ ,$$

where the integral has to be understood in the sense of isometric extension analogous to the considerations of the Fourier transform.

<u>Remark 2.6</u>: *Using that*

$$j : H \longrightarrow L_2(\mathbb{R}^+)$$
$$u \longrightarrow t^{1/2}\, u(t)$$

is a unitary mapping, we could obtain a similar result as in Theorem 2.3 for the differential expression

$$j\, \mathfrak{a}_\nu\, j^* = (2\pi D)^2 + (\nu^2 - \tfrac{1}{4})\, m^{-2},$$

compare [3], p. 544-547, [40], p. 199-205. *Then*

$$\mathcal{H}_\nu = H_\nu \circ j^* : L_2(\mathbb{R}^+) \longrightarrow L_2(\mathbb{R}^+),$$

becomes a spectral representation for $j\, A\, j^*$, *which is a selfadjoint realization of* $j\, \mathfrak{a}_\nu\, j^*$. *Note that* $j\circ k\circ \mathcal{H}_\nu \equiv j\circ Ha_\nu \circ j^* : L_2(\mathbb{R}^+) \longrightarrow L_2(\mathbb{R}^+)$ *is a Fourier integral operator (see 1.4) also referred to as <u>Hankel transform</u>, (see e.g.* [23], *p. 397). Since Fourier integral operators are associated with functions of* D_* *(see section 1.4) and since the spectral representation of* D_* *(the Mellin transform) is closely related to the Fourier transform there is an easy possibility to extend to generalized functions. A discussion of generalization to distributions based on different ideas can be found in* [41].

2.4 The Lebedev transform as a spectral representation related to the differential expression $m\,2\pi D\, m\,2\pi D + m^2$

The differential expression $a := (2\pi\ m\ D)^2 + m^2$ appears in the same connection as the one discussed in 2.3 . In fact

$$m^2(a_\nu + \lambda) = (2\pi\ m\ D)^2 + \lambda\ x^2 + \nu^2\ ,\ \nu \geq 0,\ \lambda \in \mathbb{R}. \tag{2.94}$$

The case $\lambda = 0$, $(\nu = 0)$, leads back to the differential expression already included in the discussion of paragraph 1.4 :

$$(2\pi\ m\ D)^2 = 4\ \pi^2\ (m\ D)^2.$$

The term $+\nu^2$ merely is a shift in the spectrum of $(2\pi\ m\ D)^2$, so that the spectral theory can be obtained by the same technique used for the one-sided Fourier cosine and Fourier sine transform (compare sections 1.1.3.1 and 2.2).

The case $\lambda > 0$ can be normalized to $\lambda = 1$ by the substitution

$$\mathbb{R}^+ \longrightarrow \mathbb{R}^+,$$

$$x \longrightarrow \lambda^{-1/2}\ x$$

and (2.94) goes over into

$$x^2(a_\nu + 1) = a + \nu^2,\ \nu \geq 0, \tag{2.95}$$

which is the differential expression in question. The case $\lambda < 0$ shall be briefly discussed at the end of this section.

Starting point of our investigation is - as usual - to write a in the form $b^*\ b$. An elementary calculation shows that with $\rho \equiv I_0$ (compare 2.3) and

$$b = 2\pi\ m\ \rho\ D\ \rho^{-1}, \tag{2.96}$$

we have indeed

$$a = b^*\ b,$$

where $b^* = m\ \rho^{-1}\ (2\pi D)\ \rho$ is the formal adjoint of b with respect to the inner product of $H := m^{1/2}\ L_2(\mathbb{R}^+)$. Note that $1 > I_o(x) > e^{-x}$ for $x \in \mathbb{R}^+$ (see [23], p. 84). Defining the operator

$$C : D(C) \subset m^{1/2}\ L_2(\mathbb{R}^+) \longrightarrow m^{1/2}\ L_2(\mathbb{R}^+),$$

as the closure of $\mathscr{b}^*$ defined on $C_\infty(\mathbb{R}^+)$, we are lead to consider the selfadjoint operator $A \equiv A_N = C\, C^* : H \longrightarrow H$. An element in the ortho-complement of D(C) in

$$D(B^*) \equiv \{\Phi \in H \mid \mathscr{b}^* \Phi \in H\}, \tag{2.97}$$

with respect to the graph inner product of B^* has to be a solution u of

$$(\mathscr{b}\, \mathscr{b}^* + 1)\, u = 0.$$

We may take $i\mathscr{b}\, I_1 = m\, I_0 - I_1 - I_1^2/I_0$, $-i\mathscr{b}\, K_1 = m\, K_0 + K_1 + I_1\, K_1/I_0$ as linearly independent candidates. In the evaluation of $i\, \mathscr{b}\, I_1$, $-i\mathscr{b}\, K_1$ we have made use of the relationships

$$\begin{aligned} I_{\nu-1} + I_{\nu+1} &= 2\, I'_\nu\,, \\ K_{\nu-1} + K_{\nu+1} &= -\, 2\, K'_\nu\,, \\ (m^\nu\, I_\nu)' &= m^\nu\, I_{\nu+1}, \\ (m^\nu\, K_\nu)' &= -\, m^\nu\, K_{\nu+1}, \end{aligned} \tag{2.98}$$

(see [23], p. 67). In addition to (2.81), (2.82) we quote the following asymptotic result:

$$K_\nu(x) \sim e^{-x}\, x^{-1/2} \text{ as } x \longrightarrow \infty, \tag{2.99}$$

(see [23], p. 66, p. 139).

The asymptotic behaviour of I_0, I_1, K_0, K_1 for large argument leaves us with $-i\, \mathscr{b}\, K_1$ as possible solution. But (2.82) and (2.99) show that

$$\mathscr{b}\, K_1 \notin m^{1/2}\, L_2(\mathbb{R}^+).$$

Thus, in particular,

$$C = B^*. \tag{2.100}$$

This implies

$$D(A) = \{\Phi \in H \mid \mathscr{b}\, \Phi \in H\,,\ \mathscr{a}\Phi \in H\} \tag{2.101}$$

(application of $\mathscr{a}$, $\mathscr{b}$ in the weak sense, i.e. in the sense of the definition of A^*, C^*).

Green's function ($\lambda \in \mathbb{C}\backslash\mathbb{R}$) is given by

$$G_\lambda(x,t) = \begin{Bmatrix} I_{-i\sqrt{\lambda}}(x) & K_{i\sqrt{\lambda}}(t) & \text{if} & 0 \le x \le t \\ I_{-i\sqrt{\lambda}}(t) & K_{i\sqrt{\lambda}}(x) & \text{if} & 0 \le t \le x \end{Bmatrix}, \quad x,t \in \mathbb{R}^+,$$

and shall be used (as in the previous chapter) to represent the resolvent of A .

Via Stone's formula we obtain $(0 \le \mu \le \lambda)$

$$((\Pi(\lambda) - \Pi(\mu))f \ , \ g) = \int_\mu^\lambda \overline{L(f)(\sigma)}\, L(g)(\sigma)\, d\sigma \tag{2.102}$$

$f,g \in H$, $\operatorname{supp} f \cup \operatorname{supp} g \subset\subset \mathbb{R}^+$, with

$$L(f)(\sigma) := \pi^{-1}(\sinh \pi\sqrt{\sigma})^{1/2} \int_{\mathbb{R}^+} y^{-1}\ f(x)\ K_{i(\sqrt{\sigma})}(y)\ dy \ ,$$

(observe $K_{-\nu}(z) = K_\nu(z)$ and $K_\nu(z) = \frac{\pi}{2}(\sin \pi\nu)^{-1}[I_{-\nu}(z) - I_\nu(z)]$, $\nu \notin \mathbb{Z}$).

By (2.102) L may be extended to all of H such that

$$L : H \longrightarrow L_2(\mathbb{R}^+)$$

is an isometry.

<u>Theorem 2.4</u>: A *is a selfadjoint operator with simple spectrum*

$$\sigma(A) = C\,\sigma(A) = \overline{\mathbb{R}^+} \ .$$

L *is a spectral representation for* A .

<u>Proof</u>: The first part of the theorem is clear from the considerations above and the facts that $\mathbb{R}^-$ is in the resolvent set and zero is not an eigenvalue (by (2.99)).

For the second part it remains to show that L is onto. Let $g \in N(L^*)$ and $\chi_{(a,b)}$ denote the characteristic function of the interval (a,b) , $0 \le a \le b$, then if

$$L((\Pi(b) - \Pi(a))\Phi)(\lambda) = \chi_{(a,b)}(\lambda)L(\Phi)(\lambda) \ , \quad \Phi \in H \ , \tag{2.103}$$

we have

$$\begin{aligned} 0 &= ((\Pi(b) - \Pi(a))f \ , \ L^*g) \ , \\ &= (L((\Pi(b) - \Pi(a))f) \ , \ g) \ , \\ &= (\chi_{(a,b)}Lf \ , \ g) \ , \end{aligned}$$

$$= (Lf\ ,\ \chi_{(a,b)}g)\ ,$$

in other words

$$\chi_{(a,b)}g \in N(L^*)\ . \tag{2.104}$$

This may be written as

$$0 = (L^*\ \chi_{(a,b)}g)(x) = \pi^{-1} \int_a^b (\sinh \pi\sqrt{\sigma})^{1/2}\ K_{i\sqrt{\sigma}}(x)\ g(\sigma)\ d\sigma\ , \tag{2.105}$$

(note that $K_{i\nu}(x)$, $\nu \in \mathbb{R}$, is real, [23], p. 85).

Since a,b , $0 \leq a \leq b$, are arbitrary and $(\sinh \pi\sqrt{\sigma})^{1/2}\ K_{i\sqrt{\sigma}}(x)$ is analytic in $\sqrt{\sigma}$ (and therefore the set of zeros is a Lebesgue null set) we have

$$\int_a^b g(\sigma)\ d\sigma = 0 \quad \text{for all } a,b \in \mathbb{R}^+,\ a \leq b\ ,$$

which implies

$$g = 0 \quad \text{a.e.}$$

It remains to see (2.103) . Since by partial integration we find

$$L(A\Phi)(\lambda) = \lambda\ L(\Phi)(\lambda)\ , \tag{2.106}$$

for $\Phi \in C_\infty(\mathbb{R}^+)$, a closure argument yields that (2.106) is valid for all $\Phi \in D(A)$ ($C_\infty(\mathbb{R}^+)$ is a core of A). From this we have in fact by a standard approximation procedure equation (2.103) (see appendix, Corollary A.14). This way we have indeed shown that L is onto ($L_2(\mathbb{R}^+) = N(L^*) \oplus W(L)$). □

The integral transform

$$\text{Leb} = L^* \circ k, \tag{2.107}$$

where $k : (m \sinh(\pi m))^{1/2}\ L_2(\mathbb{R}^+) \longrightarrow L_2(\mathbb{R}^+)$

$$(k\ h)(t) = \frac{\pi}{2} \frac{h(\sqrt{t})}{\sqrt{t}(\sinh \pi\sqrt{t})^{1/2}}\ ,\ t \in \mathbb{R}^+,$$

is also a unitary mapping and is known as <u>Lebedev transform</u> (see [23], p. 398, [19]). The Lebedev transform is apparently unitary as a mapping from $(m \sinh(\pi m))^{1/2}\ L_2(\mathbb{R}^+)$ onto H .

Remark 2.7: *As has been pointed out in* [20] *the limit* $U_\lambda^\pm = (A - (\lambda \pm i0+))^{-1} f$, *with (e.g.)* $f \in C_\infty(\mathbb{R}^+)$ *in — say —* $L_2^{loc}(\mathbb{R}^+)$ *can be characterized by a 'radiation condition' (at zero) of the type*

$$\ell\, U_\lambda^\pm \pm \sqrt{\lambda}\, U_\lambda^\pm \in H.$$

Similar conditions could be formulated in other cases as well. 'Radiation conditions' are not required in our approach. We only mention them in this particular context, since we encounter here the unusual situation that such a condition distinguishes U_λ^+, U_λ^- *with respect to their behaviour at a finite point rather than at infinity.*

We conclude this section with a brief look at the case $\lambda < 0$. This case can be normalized to

$$\tilde{a} := (2\pi\, m\, D)^2 - m^2.$$

Here we have

$$\tilde{\ell} = 2\pi\, m\, \tilde{\rho}\, D\, \tilde{\rho}^{-1}, \tag{2.108}$$

and

$$\tilde{\ell}^* = m\, \tilde{\rho}^{-1}\, (2\pi D)\, \tilde{\rho}, \tag{2.109}$$

where

$$\tilde{\rho} = J_0\ . \tag{2.110}$$

The differential expression $\tilde{\ell}^*$ is the formal adjoint of $\tilde{\ell}$ with respect to the inner product of $H := m^{1/2} L_2(\mathbb{R}^+)$. We have similarly to the above that $1 > J_0(x) > 0$ for $x \in \mathbb{R}^+$ (see [23], p. 146). We define — as usual — the operator

$$\underset{\sim}{C} : D(C) \subset m^{1/2} L_2(\mathbb{R}^+) \longrightarrow m^{1/2} L_2(\mathbb{R}^+),$$

as the closure of $\tilde{\ell}^*$ defined on $C_\infty(\mathbb{R}^+)$, we are lead to consider the selfadjoint operator $\tilde{A} \equiv \tilde{A}_N = \underset{\sim}{C}\, \underset{\sim}{C}^* : H \longrightarrow H$. Considering the ortho-complement of D(C) in

$$D(\underset{\sim}{B}^*) \equiv \{\Phi \in H \mid \tilde{\ell}^*\, \Phi \in H,\ x^{1/2}\, \Phi(x) \longrightarrow 0 \text{ as } x \longrightarrow \infty\}, \tag{2.111}$$

with respect to the graph inner product of $\underset{\sim}{B}^*$ we obtain

$$D(C_{\sim}) = \{\Phi \in D(B^*_{\sim}) \mid x^{1/2}\,\Phi(x) \longrightarrow 0 \quad \text{as } x \longrightarrow \infty\}. \tag{2.112}$$

It is interesting to note that in the case under discussion we obtain a boundary condition at infinity. Equation (2.111) implies

$$D(\tilde{A}) =$$

$$= \{\Phi \in H \mid \tilde{\ell}\,\Phi \in H,\ \tilde{a}\Phi \in H,\ x^{1/2}(\tilde{\ell}\,\Phi)(x) \longrightarrow 0 \text{ as } x \longrightarrow \infty\}. \tag{2.113}$$

Green's function ($\lambda \in \mathbb{C}\backslash\mathbb{R}$) is given by

$$\tilde{G}_\lambda(x,t) = \frac{i\,\pi}{2\sinh(\pi\sqrt{\lambda})}\begin{Bmatrix} J_{-i\sqrt{\lambda}}(x) & (J_{i\sqrt{\lambda}}(t) + J_{-i\sqrt{\lambda}}(t)) \text{ if } 0 \le x \le t \\ J_{-i\sqrt{\lambda}}(t) & (J_{i\sqrt{\lambda}}(x) + J_{-i\sqrt{\lambda}}(x)) \text{ if } 0 \le t \le x \end{Bmatrix},$$

for $x,t \in \mathbb{R}^+$. From this we obtain by the same reasoning as in the case $\lambda > 0$ the result:

Theorem 2.5: $\tilde{A}$ *is a selfadjoint operator with simple spectrum*

$$\sigma(\tilde{A}) = C\,\sigma(\tilde{A}) = \overline{\mathbb{R}^+}\,.$$

The unitary operator $\tilde{L}: H \longrightarrow L_2(\mathbb{R}^+)$ *defined as the isometric extension of the integral operator*

$$\tilde{L}(\Phi)(\lambda) = \frac{1}{2}\,(\operatorname{cosech}(\pi\sqrt{\lambda}))^{1/2}\int_{\mathbb{R}^+}(J_{i\sqrt{\lambda}}(x) + J_{-i\sqrt{\lambda}}(x))\,\Phi(x)\,x^{-1}\,dx,$$

for $\Phi \in C_\infty(\mathbb{R}^+)$, $\lambda \in \mathbb{R}^+$, *is a spectral representation for* $\tilde{A}$.

Compare [23], p. 150, for this particular spectral representation. The adjoint $\tilde{L}^*$ is easily calculated as

$$\tilde{L}^*(\Phi)(x) = \frac{1}{2}\int_{\mathbb{R}^+}(\operatorname{cosech}(\pi\sqrt{\lambda}))^{1/2}\,(J_{i\sqrt{\lambda}}(x) + J_{-i\sqrt{\lambda}}(x))\,\Phi(\lambda)\,d\lambda \tag{2.114}$$

for $\Phi \in C_\infty(\mathbb{R}^+)$, $x \in \mathbb{R}^+$.

2.5 The (generalized) Mehler transform as a spectral representation related to the differential expression $2\pi D\,(m^2-1)\,2\pi D + \mu^2\,(m^2-1)^{-1} - 1/4,\ \mu \in \mathbb{R}$

The differential expression $a_\mu := (2\pi D)\ (m^2 - 1)\ (2\pi D) + \mu^2\ (m^2 - 1)^{-1} - 1/4$, $\mu \geq 0$ without loss of generality, appears in the separation of variable approach to the three-dimensional Laplacian in polar coordinates. Again we start by rewriting a_μ as $\ell^* \ell$ where

$$\ell = (m^2 - 1)\ \rho\ (2\pi D)\ \rho^{-1}, \tag{2.115}$$

and ℓ^* the formal adjoint

$$\ell^* = \rho^{-1}\ (2\pi D)\ \rho, \tag{2.116}$$

with respect to $L_2((1,\infty))$ as initial space and $(m^2 - 1)^{1/2}\ L_2((1,\infty))$ as terminal space of ℓ; here $\rho = \mathcal{P}_{-1/2}^{-\mu}$. Note that as a consequence of a integral representation of $\mathcal{P}_{-1/2}^{-\mu}$ ([23], p. 184) we have

$$\begin{aligned} c_o(x + (x^2-1)^{1/2})^{-1/2-\mu} &\geq 2^\mu\ \pi^{1/2}\ \Gamma(1/2 + \mu)\ (x^2-1)^{-\mu/2}\ \mathcal{P}_{-1/2}^{-\mu}(x) \geq \\ &\geq c_o\ (x - (x^2-1)^{1/2})^{-1/2-\mu},\ x \in \mathbb{R}^+, \end{aligned} \tag{2.117}$$

with

$$c_o = \int_{(0,\pi)} (\sin\ t)^{2\mu}\ dt,$$

so that

$$\infty > \mathcal{P}_{-1/2}^{-\mu}(x) > 0 \text{ for } x \in \mathbb{R}^+. \tag{2.118}$$

We consider $A_\mu \equiv A_{N,\mu} := C_\mu\ C_\mu^*$ where C_μ is by definition the closure of ℓ^* applied to $C_\infty((1,\infty))$. For $u \in D(B_\mu^*)$ we have the following Sobolev type inequality

$$|(\rho\ u)(s) - (\rho u)(t)|^2 \leq \int_{(s,t)} |\rho(x)|^2\ dx\ \int_{(s,t)} |\rho(x)^{-1}\ (\rho(x)\ u(x))'|^2\ dx,$$

for $t \geq s > 0$. This estimate implies (according to (2.121) below) that for $\Phi \in D(C_\mu)$ the boundary condition

$$(x^2-1)^{-1/2} \Phi(x) \longrightarrow 0 \text{ as } x \longrightarrow 1+, \tag{2.119}$$

is satisfied. Here we have made use of the asymptotic result corresponding to (2.26), i.e.

$$\mathcal{P}_\nu^{-\mu}(x) \sim (x-1)^{\mu/2}.$$

To characterize $D(C_\mu)$ let us consider the ortho-complement $D(C_\mu)^\perp$ of $D(C_\mu)$ in the subspace

$$\{u \in D(B_\mu^*) \mid (x^2-1)^{-1/2} u(x) \longrightarrow 0 \text{ as } x \longrightarrow 1+\}, \tag{2.120}$$

of $D(B_\mu^*)$. Again we find that any $u \in D(C_\mu)^\perp$ must satisfy

$$(\ell\, \ell^* + 1)u = 0.$$

Candidates for solutions are

$$\ell\, \mathcal{P}_{1/2}^{-\mu} \text{ and } \ell\, \mathcal{Q}_{1/2}^{\mu}.$$

We quote the asymptotic properties of the Legendre functions $\mathcal{P}_\nu^{-\mu}$ and $\mathcal{Q}_\nu^{\mu}$ from [23], p. 196-197, [25], p. 173:

$$\begin{aligned}
&\mathcal{P}_\nu^{-\mu}(z)\ (z-1)^{-\mu/2} \longrightarrow 2^{-\mu/2}/\Gamma(\mu+1),\ \mu \geq 0,\\
&\mathcal{Q}_\nu^{\mu}(z)\ (z-1)^{\mu/2} \longrightarrow e^{i\pi\mu}\, 2^{\mu/2-1}\, \Gamma(\mu),\ \mu > 0,\\
&\mathcal{Q}_\nu^{0}(z)/\log(z-1) \longrightarrow -1/2,\ \nu \notin -\mathbb{N},\\
&\mathcal{Q}_\nu^{0\,\prime}(z) \sim (z-1)^{-1},\ \nu \notin -\mathbb{N},
\end{aligned} \tag{2.121}$$

as $z \longrightarrow 1$ and

$$\begin{aligned}
&\mathcal{P}_\nu^{-\mu}(z) \sim z^{\nu},\ \text{if } \operatorname{Re}\nu > -\tfrac{1}{2},\\
&\mathcal{P}_\nu^{-\mu}(z) \sim z^{-\nu-1},\ \text{if } \operatorname{Re}\nu < -\tfrac{1}{2},\\
&\mathcal{P}_{-1/2}^{-\mu}(z) \sim z^{-1/2} \ln z,\\
&\mathcal{Q}_\nu^{\mu}(z) \sim z^{-\nu-1},\ \text{if } -\nu \notin \mathbb{N} + \tfrac{1}{2},\\
&\mathcal{Q}_\nu^{\mu}(z) \sim z^{-\nu-2\mu-3},\ \text{if } -\nu \in \mathbb{N} + \tfrac{1}{2},
\end{aligned} \tag{2.122}$$

as $|z| \longrightarrow \infty$, $z \in \mathbb{C}-(-\infty,-1)$. We have from the asymptotic behaviour of $\mathcal{P}_\nu^{-\mu}$ and $\mathcal{Q}_\nu^{\mu}$ that

$$\mathcal{P}_\nu^{-\mu}(x)\ \mathcal{Q}_\sigma^{\mu}(x) \longrightarrow e^{i\pi\mu}/(2\mu),\ \mu > 0,\ \nu,\sigma \in \mathbb{C},$$
$$\mathcal{P}_\nu^{0}(x)\ \mathcal{Q}_\sigma^{0}(x)/\ \ln(x-1) \longrightarrow -1/2,\ \nu,\ \sigma \in \mathbb{C},\ \sigma \notin -\mathbb{N}, \tag{2.123}$$

as $x \longrightarrow 1+$. To calculate $\ell\, \mathcal{P}_{1/2}^{-\mu}$ and $\ell\, \mathcal{Q}_{1/2}^{\mu}$ we observe

$$(m^2-1)\ 2\pi i\ D\ \mathcal{P}_\nu^{-\mu} = (\mu+\nu+1)\ \mathcal{P}_{\nu+1}^{-\mu} - (\nu+1)\ m\ \mathcal{P}_\nu^{-\mu},$$
$$= \nu\ m\ \mathcal{P}_\nu^{-\mu} - (\nu-\mu)\ \mathcal{P}_{\nu-1}^{-\mu},$$
$$(m^2-1)\ 2\pi i\ D\ \mathcal{Q}_\nu^{\mu} = (-\mu+\nu+1)\ \mathcal{Q}_{\nu+1}^{\mu} - (\nu+1)\ m\ \mathcal{Q}_\nu^{\mu},$$
$$= \nu\ m\ \mathcal{Q}_\nu^{\mu} - (\nu+\mu)\ \mathcal{Q}_{\nu-1}^{\mu}, \tag{2.124}$$

(recall that m denotes the formal operator of multiplication by the argument). Using (2.124) we obtain

$$\ell\, \mathcal{P}_{1/2}^{-\mu} = i\ ((1/2+\mu)\ (\mathcal{P}_{1/2}^{-\mu})^2/\mathcal{P}_{-1/2}^{-\mu} + (1/2-\mu)\ \mathcal{P}_{-1/2}^{-\mu} - m\ \mathcal{P}_{1/2}^{-\mu},$$
$$\ell\, \mathcal{Q}_{1/2}^{\mu} = i\ ((1/2+\mu)\ \mathcal{P}_{1/2}^{-\mu}\ \mathcal{Q}_{1/2}^{-\mu}/\mathcal{P}_{-1/2}^{-\mu} + (1/2+\mu)\ \mathcal{Q}_{-1/2}^{-\mu} - m\ \mathcal{Q}_{1/2}^{-\mu}).$$

According to (2.122) only $\ell\, \mathcal{Q}_{1/2}^{\mu}$ is well enough behaved at ∞ to be an element of

$$(m^2-1)^{1/2}\ L_2((1,\infty)).$$

However, according to (2.123) the function $\ell\, \mathcal{Q}_{1/2}^{\mu}$ satisfies

$$\ell\, \mathcal{Q}_{1/2}^{\mu}(x) \sim (x-1)^{-\mu/2},\ \mu > 0,$$
$$\ell\, \mathcal{Q}_{1/2}^{0}(x) \sim \ln(x-1),$$

and, thus, does not satisfy the boundary condition (2.119). Consequently we have

$$D(C_\mu) = \{u \in D(B_\mu^*) \mid (x^2-1)^{-1/2}\, u(x) \longrightarrow 0 \text{ as } x \longrightarrow 1+\}. \tag{2.125}$$

The domain of A_μ is therefore characterized by

$$D(A_\mu) = \{\Phi \in L_2((1,\infty)) \mid (m^2-1)^{-1/2}\, \ell\, \Phi \in L_2((1,\infty)),$$

$$\lim_{x \longrightarrow 1+} (x^2-1)^{-1/2}\, u(x) = 0,\ \mathcal{a}_\mu \Phi \in L_2((1,\infty))\}.$$

Another look at (2.121), (2.122) reveals that the deficiency indices are $(0|0)$ for $\mu \geq 1$ and that they are $(1|1)$ if $\mu \in [0,1)$. To see this observe that $\mathcal{P}_{\nu_\pm}^{-\mu}$, $\mathcal{Q}_{\nu_\pm}^{\mu}$ with $\nu_\pm = -1/2 + 2^{-1/2}(1 \mp i))$ are possible candidates for being in $N(A_\mu^* \pm i)$. In the latter case we have

$$N(A_\mu^* \pm i) = \operatorname{span} \{\mathcal{Q}_{\nu_\pm}^{\mu}\}. \tag{2.126}$$

This consideration shows that only for $\mu \in [0,1)$ we need a boundary condition to describe $D(A_\mu)$. Moreover, for $\mu \geq 1$ we have

$$D(A_\mu) = D(A_\mu) =$$

$$= \{\Phi \in L_2((1,\infty)) \mid (m^2-1)^{-1/2} \ell\Phi \in L_2((1,\infty)),\ \mathcal{a}_\mu \Phi \in L_2((1,\infty))\}. \tag{2.127}$$

In the same manner as in previous paragraphs we represent the resolvent using the corresponding Green function ($\lambda \in \mathbb{C}\setminus\mathbb{R}$)

$$G_{\mu,\lambda}(x,t) = e^{-i\pi\mu} \begin{cases} \mathcal{P}_\nu^{-\mu}(x)\ \mathcal{Q}_\nu^{\mu}(t) & \text{for } 0 \leq x \leq t \\ \mathcal{P}_\nu^{-\mu}(t)\ \mathcal{Q}_\nu^{\mu}(x) & \text{for } 0 \leq t \leq x \end{cases}, \tag{2.128}$$

where

$$\nu = -\frac{1}{2} - i\sqrt{\lambda}\ .$$

By Stone's formula we obtain for $b \geq a \geq 0$ a representation of the spectral family $(\Pi(\lambda))_{\lambda \in \mathbb{R}}$ of A_μ :

$$(\Pi(b) - \Pi(a)f\ ,\ g) = \int_a^b \overline{M^\mu(f)(\sigma)}\ M^\mu(g)(\sigma)\ d\sigma$$

$f, g \in L_2((1,\infty))$, $\operatorname{supp} f \cup \operatorname{supp} g \subset\subset (1,\infty)$, where

$$M^{\mu}(f)(\sigma) = (2\pi)^{-1/2} (\sinh \pi\sqrt{\sigma})^{1/2} \Gamma(\tfrac{1}{2} + i\sqrt{\sigma} + \mu) \cdot$$

$$\cdot \int_{(1,\infty)} \mathscr{P}^{-\mu}_{-1/2 + i\sqrt{\sigma}}(t) f(t)\, dt$$

(note: $\mathscr{Q}^{\mu}_{-\nu-1}(z) - \mathscr{Q}^{\mu}_{\nu}(z) = e^{i\pi\mu} \cos(\pi\nu)\, \Gamma(\nu+\mu+1)\, \Gamma(\mu-\nu) \mathscr{P}^{-\mu}_{\nu}(z)$, [23], p 164).

The isometry M^{μ} can be extended to all of $L_2((1,\infty))$ by continuity. By the same arguments as in the proof of Theorem 2.4 we can show

<u>Theorem 2.6</u>: *The selfadjoint operator* A_{μ} , $\mu \geq 0$, *has a simple spectrum*

$$\sigma(A_{\mu}) = C\, \sigma(A_{\mu}) = \overline{\mathbb{R}^{+}} .$$

The integral transform $M^{\mu}: L_2((1,\infty)) \to L_2(\mathbb{R}^{+})$ *is a spectral representation for* A_{μ} .

<u>Proof</u>: Since the arguments are indeed the same as in paragraph 2.4 we leave the details as an exercise and assume everything is proved if we can show the analogue of (2.106) .

As in the proof of theorem 2.4 we see that

$$M^{\mu}(A_{\mu}\Phi)(\lambda) = \lambda\, M^{\mu}(\Phi)(\lambda) \tag{2.129}$$

for $\Phi \in C_{\infty}((1,\infty))$. Since $C_{\infty}((1,\infty))$ is dense in $D(A_{\mu})$ for $\mu \geq 1$, we are left with $\mu \in [0,1)$ as a case to consider. Since in this case we have (2.126) it follows from the general theory of selfadjoint extensions (compare (1.15) and Appendix) that

$$D(A_{\mu}) = D(A_{\mu}) \oplus \Gamma_{\mu} ,$$

where

$$\Gamma_{\mu} = \mathrm{span}\{ \mathscr{Q}^{\mu}_{\nu_-} + \alpha\, \mathscr{Q}^{\mu}_{\nu_+} \}$$

for some appropriate $\alpha \in \mathbb{C}$, $|\alpha| = 1$. Because of (2.121) only the choice $\alpha = -1$ leads to

$$\Gamma_{\mu} \subset D(A_{\mu}) .$$

Thus, we only have to show that (2.129) is valid for

$$\Phi = \mathcal{Q}^{\mu}_{\nu_-} - \mathcal{Q}^{\mu}_{\nu_+} .$$

But (2.129) follows for this particular Φ by partial integration likewise. Vanishing of the boundary terms may be deduced from the asymptotic behaviour of $\mathcal{Q}^{\mu}_{\nu}$, $\mathcal{P}^{-\mu}_{\nu}$ in boundary point +1 and at ∞ (see (2.121) and (2.122)). This completes the proof of the theorem.

□

The adjoint of M^{μ} can be written in integral forms as

$$(M^{\mu})*(g)(x) =$$

$$= (2\pi)^{-1/2} \int_{\mathbb{R}^+} (\sinh \pi\sqrt{\lambda})^{1/2}\, \Gamma(\tfrac{1}{2} - i\sqrt{\lambda}+\mu)\, \mathcal{P}^{-\mu}_{-1/2+i\sqrt{\lambda}}(x)\, g(\lambda)\, d\lambda \qquad (2.130)$$

$$= (2\pi)^{-1/2} \int_{\mathbb{R}^+} (\sinh \pi t)^{1/2}\, t\, \Gamma(\tfrac{1}{2} - it+\mu)\, \mathcal{P}^{-\mu}_{-1/2+it}(x)\, g(t^2)\, dt .$$

Thus, if we introduce the unitary mapping

$$k : L_2(\mathbb{R}^+) \longrightarrow \pi^{-1/2} (\sinh \pi m)^{1/2}\, m^{1/2}\, L_2(\mathbb{R}^+),$$

$$g \longrightarrow (\tfrac{2}{\pi})^{1/2} (\sinh \pi m)^{1/2}\, m\, g(m^2) ,$$

we have an integral transform

$$\mathrm{Meh}_{\mu} = (M^{\mu})^{*} k^{*} \qquad (2.131)$$

called <u>(generalized) Mehler transform</u>. It is

$$\mathrm{Meh}_{\mu}(f)(x) = \int_{\mathbb{R}^+} \mathcal{P}^{-\mu}_{-1/2+i\sqrt{\lambda}}(x)\, f(t)\, dt . \qquad (2.132)$$

where the integral has to be understood in the sense that

$$\mathrm{Meh}_{\mu} : \pi^{-1/2} (\sinh \pi m)^{1/2}\, m^{1/2}\, L_2(\mathbb{R}^+) \longrightarrow L_2((1,\infty))$$

is unitary.

<u>Remark 2.8</u>: *The Mehler transform is usually written with* $-\mu$ *replaced by* μ. *We have assumed that in the above formulae* μ *should be non-negative. In this context it is interesting to note that* M^{μ} *is still a spectral representation for an operator related to* $\mathfrak{a}_{\mu}$ *(by a proper choice of boundary condition) if* $\mu \in (-1,0)$. *Furthermore, since*

$$\mathcal{P}_\nu^{-m} = \frac{\Gamma(\nu-m+1)}{\Gamma(\nu+m+1)} \mathcal{P}_\nu^m , \quad m \in \mathbf{Z} ,$$

we may extend the definition of M^μ

$$M^m = M^{-m} , \quad m \in \mathbf{Z} .$$

2.6 The Jacobi transform as a spectral representation for a selfadjoint realization of (α, ß ε IR) $(m-1)^{-\alpha}$ $(m+1)^{-}$ $2\pi D\,(m-1)^{\alpha+1}\,(m+1)^{\beta+1}\,2\pi D - (\alpha+\beta+1)$

The differential expression

$$\mathfrak{a}_{\alpha,\beta} := (m-1)^{-\alpha}\ (m+1)^{-\beta}\ 2\pi D\ (m-1)^{\alpha+1}\ (m+1)^{\beta+1}\ 2\pi D\ -\ (\alpha+\beta+1)^2/4,$$

is apparently closely related to the one discussed in section 2.1.3. In fact there is only a change in sign (since we are now interested in the semi-infinite interval $(1,\infty)$) and a shift in the spectral parameter (by $(\alpha+\beta+1)^2/4$). Accordingly the solutions are very similar. We have as linearly independent solutions of

$$(\mathfrak{a}_{\alpha,\beta} - \lambda)\ u = 0, \qquad (2.133)$$

in the case $\alpha \notin \mathbb{Z}$

$$f_\lambda^{(\alpha,\beta)} = F(\vartheta - i\sqrt{\lambda}\ ,\ \vartheta + i\sqrt{\lambda}\ ,\ \alpha + 1;\ \frac{1-m}{2})/\Gamma(\alpha+1),$$

$$g_\lambda^{(\alpha,\beta)} = \left(\frac{m-1}{2}\right)^{-\alpha} F(1-\tilde{\vartheta}-i\sqrt{\lambda},\ 1-\tilde{\vartheta}+i\sqrt{\lambda},\ 1-\alpha;\ \frac{1-m}{2})/\Gamma(1-\alpha), \qquad (2.134)$$

where $\sqrt{\lambda}$ is chosen such that $\mathrm{Im}\,\sqrt{\lambda} \geq 0$ and

$$\vartheta = \frac{\alpha+\beta+1}{2}\ ,\quad \tilde{\vartheta} = \frac{\alpha-\beta+1}{2}.$$

($F(a,b,c;\ z)$ is the standard principal solution of the hypergeometric equation). Linear independence can be seen from the Wronskian

$$W(g_\lambda^{(\alpha,\beta)}, f_\lambda^{(\alpha,\beta)}) = \frac{\sin(\pi\alpha)}{2\pi}\ \left(\frac{m+1}{2}\right)^{-\beta-1}\ \left(\frac{m-1}{2}\right)^{-\alpha-1},$$

For the time being we shall assume that $\alpha \notin \mathbb{Z}$. To give $\mathfrak{a}_{\alpha,\beta}$ the standard form $\mathfrak{b}^*_{\alpha,\beta}\ \mathfrak{b}_{\alpha,\beta}$ we define

$$\mathfrak{b}_{\alpha,\beta} = 2\pi\ (m-1)^{\alpha+1}\ (m+1)^{\beta+1}\ \rho\ D\ \rho^{-1},$$

$$\mathfrak{b}^*_{\alpha,\beta} = 2\pi\ (m-1)^{-\alpha}\ (m+1)^{-\beta}\ \rho^{-1}\ D\ \rho, \qquad (2.135)$$

where

$$\rho = F(\vartheta,\vartheta,\alpha+1;\ \frac{1-m}{2}).$$

Note that by Euler's integral representation for hypergeometric functions $F(\vartheta,\vartheta,\alpha+1;\ \frac{1-m}{2})$ is positive and bounded in $(1,\infty)$. Application of Euler's representation requires

$$|\beta| < \alpha+1.$$

Since we have ([23], p. 47)

$$F(a,b,c;z) = (1-z)^{c-b-a}\, F(c-a,c-b,c;z), \tag{2.136}$$

we see that

$$F(\vartheta,\vartheta,\alpha+1;\frac{1-m}{2}) = (\frac{m+1}{2})^{-\beta}\, F(\tilde{\vartheta},\tilde{\vartheta},\alpha+1,\frac{1-m}{2}), \tag{2.137}$$

and so since

$$F(0,0,\alpha+1;\frac{1-m}{2}) = 1,$$

we may include the case $|\beta| = \alpha+1$. So, we shall assume that the condition

$$|\beta| \le \alpha + 1, \tag{2.138}$$

is satisfied. The differential expressions $\ell_{\alpha,\beta}$, $\ell^*_{\alpha,\beta}$ are formally adjoint with respect to

$$H_{\alpha,\beta} = (m-1)^{-\alpha/2}\ (m+1)^{-\beta/2}\ L_2((1,\infty)), \tag{2.139}$$

as initial space and $(m-1)^{\alpha/2+1/2}(m+1)^{\beta/2+1/2}\ L_2((1,\infty))$ as terminal space for $\ell_{\alpha,\beta}$. Recall that by our generic definition $C_{\alpha,\beta}$ denotes the closure of $\ell^*_{\alpha,\beta}$ with respect to the domain $C_\infty((1,\infty))$. Thus, $C_\infty((1,\infty))$ is dense in $D(C_{\alpha,\beta})$ with respect to the graph norm of

$$C_{\alpha,\beta} : (m-1)^{\alpha/2+1/2}(m+1)^{\beta/2+1/2}\ L_2((1,\infty)) \longrightarrow H_{\alpha,\beta} .$$

The differential expression $\ell_{\alpha,\beta}$ corresponds to the adjoint operator $C^*_{\alpha,\beta}$. Again we only consider the Neumann case, i.e.

$$A_{\alpha,\beta} \equiv A_{N,\alpha,\beta} = C_{\alpha,\beta}\ C^*_{\alpha,\beta} . \tag{2.140}$$

Let us also assume that

$$\alpha > -1. \tag{2.141}$$

Further restrictions shall be imposed as they come up as required and previous constraints may be waived by particular case studies during our investigation. The Sobolev estimate

$$|\rho(s)\ \Phi(s)\ -\ \rho(t)\ \Phi(t)|^2 \leq$$

$$\leq \int_{(s,t)} (x+1)^{\beta}\ (x-1)^{\alpha}\ \rho^2(x)\ dx \int_{(s,t)} (x+1)^{\beta}\ (x-1)^{\alpha}\ |(\ell^*_{\alpha,\beta}\ \Phi)(x)|^2\ dx,$$

for $\Phi \in D(B^*_{\alpha,\beta})$, $1 \leq s \leq t < \infty$, yields as a boundary condition for any element Φ in $D(C_{\alpha,\beta})$

$$(x-1)^{-\alpha/2-1/2}\ \Phi(x) \longrightarrow 0 \text{ as } x \longrightarrow 1+, \tag{2.142}$$

(compare (2.51)). Consider the ortho-complement $D(C_{\alpha,\beta})^{\perp}$ of subspace $D(C_{\alpha,\beta})$ in $D(B^*_{\alpha,\beta})$. An element $u \in D(C_{\alpha,\beta})^{\perp}$ has to satisfy

$$(\ell_{\alpha,\beta}\ \ell^*_{\alpha,\beta} + 1) = 0. \tag{2.143}$$

We have two linearly independent solutions

$$\ell_{\alpha,\beta}\ f^{(\alpha,\beta)}_{-1},\ \ \ell_{\alpha,\beta}\ g^{(\alpha,\beta)}_{-1}. \tag{2.144}$$

Using (2.39) we can evaluate these as

$$\ell_{\alpha,\beta}\ f^{(\alpha,\beta)}_{-1} =$$

$$= -i\ (m-1)^{\alpha+1}\ (m+1)^{\beta+1} \left\{(\vartheta^2-1)\ f^{(\alpha+1,\beta+1)}_{-1}\ -\ \vartheta^2\ f^{(\alpha,\beta)}_{-1}\ \frac{f^{(\alpha+1,\beta+1)}_0}{f^{(\alpha,\beta)}_0}\right\},$$

$$\ell_{\alpha,\beta}\ g^{(\alpha,\beta)}_{-1} =$$

$$= i\ (m-1)^{\alpha+1}\ (m+1)^{\beta+1} \left\{(m-1)^{-1}(\alpha\ g^{(\alpha,\beta)}_{-1}\ +\ ((1-\tilde{\vartheta})^2-1)\ (1-\alpha)\ g^{(\alpha-1,\beta+1)}_{-1}\ +\right.$$

$$\left. +\ \vartheta^2\ g^{(\alpha,\beta)}_{-1}\ \frac{f^{(\alpha+1,\beta+1)}_0}{f^{(\alpha,\beta)}_0}\right\}.$$

In order to proceed with our usual line of reasoning we need asymptotic properties of $f^{(\alpha,\beta)}_{\lambda}$, $g^{(\alpha,\beta)}_{\lambda}$. From the asymptotic properties of the hypergeometric function we have

$$f_\lambda^{(\alpha,\beta)}(x) \longrightarrow 1/\Gamma(1+\alpha)\ ,\ g_\lambda^{(\alpha,\beta)}(x)\ (x-1)^\alpha \longrightarrow 2^\alpha/\Gamma(1-\alpha), \qquad (2.145)$$

as $x \longrightarrow 1+$, (compare (2.40)). To obtain the asymptotic behaviour of $f_\lambda^{(\alpha,\beta)}$, $g_\lambda^{(\alpha,\beta)}$ at infinity for $\lambda \notin -\mathbb{N} \cup \{0\}$ we can use

$$F(a,b,c;z) = (1-z)^{-a} \frac{\Gamma(c)\ \Gamma(b-a)}{\Gamma(b)\ \Gamma(c-a)} F(a,c-b,a-b+1;\frac{1}{1-z}) + \\ + (1-z)^{-b} \frac{\Gamma(c)\ \Gamma(a-b)}{\Gamma(a)\ \Gamma(c-b)} F(b,c-a,b-a+1;\frac{1}{1-z}), \qquad (2.146)$$

for $z \in \mathbb{Z}$, $\arg(1-z) \in (-\pi,\pi)$, $(a-b) \notin \mathbb{Z}$, and for $\lambda \in -\mathbb{N} \cup \{0\}$

$$F(a,a+m,c;z) \sim (-z)^{-a-m} \ln(-z) \text{ as } |z| \longrightarrow \infty,\ m \in \mathbb{N} \cup \{0\}. \qquad (2.147)$$

for $z \in \mathbb{C}$ with $\arg(-z) \in (-\pi,\pi)$, $(c-a) \notin \mathbb{Z}$, (see [23], p. 48). Observing that only $\ell_{\alpha,\beta}\, f_{-1}^{(\alpha,\beta)}$ satisfies the boundary condition (2.142) and that according to (2.146), (2.147) we have for $\tilde{\vartheta} \notin \mathbb{Z}$

$$x^\vartheta = O(\ell_{\alpha,\beta}\, f_{-1}^{(\alpha,\beta)}(x)) \text{ as } x \longrightarrow \infty, \qquad (2.148)$$

thus $\ell_{\alpha,\beta}\, f_{-1}^{(\alpha,\beta)} \notin (m-1)^{\alpha/2+1/2}(m+1)^{\beta/2+1/2}\, L_2((1,\infty))$ in this case. Consequently we have

$$D(C_{\alpha,\beta}) = \\ = \{\ \Phi \in D(\overset{\circ}{B}{}^*_{\alpha,\beta})\ |\ (x-1)^{-\alpha/2-1/2}\Phi(x) \longrightarrow 0 \text{ as } x \longrightarrow 1+\}. \qquad (2.149)$$

Noticing that according to (2.145) for $\alpha > 0$

$$\ell_{\alpha,\beta}\, g_{-1}^{(\alpha,\beta)} \notin (m-1)^{\alpha/2+1/2}(m+1)^{\beta/2+1/2}\, L_2((1,\infty)),$$

we can sharpen the description of $D(C_{\alpha,\beta})$ in this case to yield

$$C_{\alpha,\beta} = B^*_{\alpha,\beta}\ ,\ \alpha > 0. \qquad (2.150)$$

Similar as in 2.1.3 we therefore have for the domain of $A_{\alpha,\beta}$

$$D(A_{\alpha,\beta}) = \{\Phi\in H_{\alpha,\beta}\ |\ (m^2-1)^{1/2}\Phi' \in H_{\alpha,\beta}\ ,\ a_{\alpha,\beta}\,\Phi \in H_{\alpha,\beta}\ , \\ (x-1)^{\alpha/2+1/2}\ \Phi'(x) \longrightarrow 0 \text{ as } x \longrightarrow 1\ \}, \qquad (2.151)$$

and for $\alpha \geq 0$

$$D(A_{\alpha,\beta}) = \{\Phi\in H_{\alpha,\beta}\ |\ (m^2-1)^{1/2}\Phi' \in H_{\alpha,\beta}\ , a_{\alpha,\beta}\,\Phi \in H_{\alpha,\beta}\ \}. \qquad (2.152)$$

We still have to consider the case $\tilde{\vartheta} \in \mathbb{Z}$. Since we have assumed that $\alpha \notin \mathbb{Z}$ we can use (2.137) to obtain the same asymptotics as found in (2.148). Thus, formulae (2.149) — (2.152) remain valid.

Now if we remove the assumption $\alpha \notin \mathbb{Z}$ we have to use a different solution in place of $g_\lambda^{(\alpha,\beta)}$ as defined above. We shall not have to go into detail. It suffices to notice that the second solution — let us call it again $g_\lambda^{(\alpha,\beta)}$ — as obtained by the substitution $z = \frac{1-x}{2}$, from [23], p. 45, has asymptotic behaviour

$$g_\lambda^{(\alpha,\beta)}(x) \sim (x-1)^{-\alpha} \text{ as } x \longrightarrow 1+, \quad \alpha \in \mathbb{N},$$

$$g_\lambda^{(0,\beta)}(x) \sim \ln(x-1) \text{ as } x \longrightarrow 1+,$$

and thusly

$$\ell_{\alpha,\beta}\, g_{-1}^{(\alpha,\beta)} \notin (m-1)^{\alpha/2+1/2}(m+1)^{\beta/2+1/2}\, L_2((1,\infty)), \quad \alpha \in \mathbb{N} \cup \{0\}. \tag{2.153}$$

(Notice that we maintain the assumption $\alpha > -1$).

We shall now determine a spectral representation of $A_{\alpha,\beta}$. To avoid lengthy case studies let us assume that $\alpha \notin \mathbb{Z}$. As before we start by representing the resolvent of $A_{\alpha,\beta}$ in terms of Green's function. Green's function for our problem is given by

$$G_\lambda^{(\alpha,\beta)}(x,y) = c\,(h(x-y)\, u_{\lambda,\alpha,\beta}(x)\, v_{\lambda,\alpha,\beta}(y) + \\ + h(y-x)\, u_{\lambda,\alpha,\beta}(y)\, v_{\lambda,\alpha,\beta}(x)),$$

for $\lambda \in \mathbb{C} - \mathbb{R}$, $x,y \in (1,\infty)$, where

$$c = \frac{\pi}{\sin(\pi\alpha)}\, 2^{-(\alpha+\beta+1)},$$

$$v_{\lambda,\alpha,\beta} = f_\lambda^{(\alpha,\beta)},$$

$$u_{\lambda,\alpha,\beta} = g_\lambda^{(\alpha,\beta)} - \omega(\sqrt{\lambda})\, f_\lambda^{(\alpha,\beta)},$$

with

$$\omega(\sqrt{\lambda}) = \frac{\Gamma(1-\vartheta-i\sqrt{\lambda})\,\Gamma(1-\tilde{\vartheta}-i\sqrt{\lambda})}{\Gamma(\vartheta-i\sqrt{\lambda})\,\Gamma(\tilde{\vartheta}-i\sqrt{\lambda})}. \tag{2.154}$$

That $v_{\lambda,\alpha,\beta}$ has this form is clear from the observation that $f_\lambda^{(\alpha,\beta)}$ is the only solution that satisfies the constraints at 1 required for a function to be in $D(C^*_{\alpha,\beta})$. To determine $u_{\lambda,\alpha,\beta}$ we observe that by (2.139) we have

$$f_\lambda^{(\alpha,\beta)} = (\tfrac{m+1}{2})^{-\vartheta+i\sqrt{\lambda}}\, c_o(\sqrt{\lambda})\; f(\sqrt{\lambda};\bullet) + (\tfrac{m+1}{2})^{-\vartheta-i\sqrt{\lambda}}\, c_o(-\sqrt{\lambda})\; f(-\sqrt{\lambda};\bullet),$$

where

$$c_o(t) = \frac{\Gamma(2it)}{\Gamma(\vartheta+it)\;\Gamma(\tilde{\vartheta}+it)},$$

$$f(t;x) = F(\vartheta-it,\tilde{\vartheta}-it,1-2it;\tfrac{2}{x+1}), \quad x \in (1,\infty).$$

Similarly we see

$$g_\lambda^{(\alpha,\beta)} = (\tfrac{m-1}{2})^{-\alpha} \Big\{ (\tfrac{m+1}{2})^{-1+\tilde{\vartheta}+i\sqrt{\lambda}}\, d_o(\sqrt{\lambda})\; g(\sqrt{\lambda};\bullet) + $$

$$+ (\tfrac{m+1}{2})^{-1+\tilde{\vartheta}-i\sqrt{\lambda}}\, d_o(-\sqrt{\lambda})\; g(-\sqrt{\lambda};\bullet) \Big\},$$

with

$$d_o(t) = \frac{\Gamma(2it)}{\Gamma(1-\vartheta+it)\;\Gamma(1-\tilde{\vartheta}+it)},$$

$$g(t;x) = F(1-\tilde{\vartheta}-it,1-\vartheta-it,1-2it;\tfrac{2}{x+1}), \quad x \in (1,\infty).$$

From this we have the asymptotic behaviour at infinity

$$f_\lambda^{(\alpha,\beta)}(x) = 2^{\vartheta-i\sqrt{\lambda}}\, x^{-\vartheta+i\sqrt{\lambda}}\, c_o(\sqrt{\lambda}) + 2^{\vartheta+i\sqrt{\lambda}}\, x^{-\vartheta-i\sqrt{\lambda}}\, c_o(-\sqrt{\lambda}) + $$

$$+ O(x^{-\vartheta-i\sqrt{\lambda}-1}),$$

$$g_\lambda^{(\alpha,\beta)}(x) = 2^{\vartheta-i\sqrt{\lambda}}\, x^{-\vartheta+i\sqrt{\lambda}}\, d_o(\sqrt{\lambda}) + 2^{\vartheta+i\sqrt{\lambda}}\, x^{-\vartheta-i\sqrt{\lambda}}\, d_o(-\sqrt{\lambda}) + $$

$$+ O(x^{-\vartheta-i\sqrt{\lambda}-1}),$$

as $x \longrightarrow \infty$. From this we have at least for small $\mathrm{Im}(\sqrt{\lambda})$

$$g_\lambda^{(\alpha,\beta)}(x) - \omega(\sqrt{\lambda})\; f_\lambda^{(\alpha,\beta)}(x) \sim x^{-\vartheta+i\sqrt{\lambda}} \text{ as } x \longrightarrow \infty, \tag{2.155}$$

which confirms our choice of $u_{\lambda,\alpha,\beta}$ (; note that by our choice of $\sqrt{\lambda}$ we have $m^{-\vartheta+i\sqrt{\lambda}} \in H_{\alpha,\beta}$). Thus we have the following representation for the resolvent of $A_{\alpha,\beta}$

$$(A_{\alpha,\beta} - \lambda)^{-1} f)(x) = c\, u_{\lambda,\alpha,\beta}(x) \int_1^x v_{\lambda,\alpha,\beta}(y)\ (y+1)^{\beta}\ (y-1)^{\alpha}\ f(y)\ dy\ +$$

$$+\ c\, v_{\lambda,\alpha,\beta}(x) \int_x^{\infty} u_{\lambda,\alpha,\beta}(y)\ (y+1)^{\beta}\ (y-1)^{\alpha}\ f(y)\ dy, \qquad (2.156)$$

for $f \in C_{\infty}((1,\infty))$, $\lambda \in \mathbb{C}-\mathbb{R}$, $|\mathrm{Im}(\sqrt{\lambda})| < 1$. For Stone's formula we only need to consider terms of the form

$$(\varphi, (A_{\alpha,\beta} - \lambda)^{-1} f) =$$

$$= c \int_{(1,\infty)} \overline{\varphi(x)}\ (x+1)^{\beta}(x-1)^{\alpha} u_{\lambda,\alpha,\beta}(x) \int_1^x v_{\lambda,\alpha,\beta}(y)\ (y+1)^{\beta}(y-1)^{\alpha} f(y)\ dy\ dx\ +$$

$$+\ c \int_{(1,\infty)} \overline{\varphi(x)}\ (x+1)^{\beta}(x-1)^{\alpha} v_{\lambda,\alpha,\beta}(x) \int_x^{\infty} u_{\lambda,\alpha,\beta}(y)\ (y+1)^{\beta}(y-1)^{\alpha} f(y)\ dy\ dx,$$

for φ, $f \in C_{\infty}((1,\infty))$, $\lambda = \lambda_0 \pm i\varepsilon$, $\varepsilon, \lambda_0 \in \mathbb{R}^{+}$, $\varepsilon < 1$. We have that

$$F(a+it, a-it, c; \tfrac{1-m}{2}) \text{ is an even function of } t \in \mathbb{R}, \qquad (2.157)$$

for $a, c \in \mathbb{R}$, $c \notin \mathbb{Z}$. This follows from the fact that the function in (2.157) solves the same equation as the one where t is replaced by -t and that both have the same asymptotic behaviour at boundary point 1. Using (2.157) we obtain from Stone's formula a representation of the spectral family $(\Pi(\lambda))_{\lambda\in\mathbb{R}}$ of $A_{\alpha,\beta}$:

$$(\Pi(b) - \Pi(a) f\ ,\ g) = \int_a^b \overline{W^{(\alpha,\beta)}(f)(\sigma)}\ W^{(\alpha,\beta)}(g)(\sigma)\ d\sigma \qquad (2.158)$$

$f, g \in L_2((1,\infty))$, $\mathrm{supp}\, f \cup \mathrm{supp}\, g \subset\subset (1,\infty)$, $b \geq a > 0$, where we may choose

$$W^{(\alpha,\beta)}(g)(\sigma) = \left(\frac{ic\omega(-\sqrt{\sigma}) + \overline{ic\omega(-\sqrt{\sigma})}}{2\pi}\right)^{1/2} \int_{(1,\infty)} f_{\sigma}^{(\alpha,\beta)}(x)\ (x+1)^{\beta}(x-1)^{\alpha} g(x)\ dx,$$

for $g \in C_{\infty}((1,\infty))$. By letting $a \longrightarrow 0$ and $b \longrightarrow \infty$ in (2.158) and by a density argument the integral transform $W^{(\alpha,\beta)}$ can be extended as an isometry to all of $H_{\alpha,\beta}$

$$W^{(\alpha,\beta)}: H_{\alpha,\beta} \longrightarrow L_2(\mathbb{R}^+).$$

(Note that 0 is not an eigenvalue !). From Stones formula we have that the expression

$$\omega_0(\sigma) = (i\ c\ \omega(-\sqrt{\sigma}) + \overline{i\ c\ \omega(-\sqrt{\sigma})})/2\pi\ ,\ \sigma \in \mathbb{R}^+, \tag{2.159}$$

is a non-negative real number. That it is a real number is clear since the gamma function Γ is a real function, i.e.

$$\overline{\Gamma(z)} = \Gamma(\bar{z}),\ z \in \mathbb{C},\ z \notin -\mathbb{N} \cup \{0\}.$$

An elementary calculation (making generous use of the relation

$$\Gamma(1-z)\ \Gamma(z) = \frac{\pi}{\sin(\pi z)}\ ,\ z \in \mathbb{C},\ z \notin \mathbb{N}),$$

yields

$$\omega_0(\sigma) = \pi^2 \frac{\sinh(2\pi\sqrt{\sigma})\quad 2^{-\alpha-\beta-2}}{|\Gamma(\vartheta+i\sqrt{\sigma})|^2\ |\Gamma(\tilde{\vartheta}+i\sqrt{\sigma})|^2\ |\sin(\pi(\vartheta+i\sqrt{\sigma}))|^2\ |\sin(\pi(\tilde{\vartheta}+i\sqrt{\sigma}))|^2}\ ,$$

$$= \pi^{-2}\ 2^{-\alpha-\beta-1}\ \sinh(\pi\sqrt{\sigma})\ \cosh(\pi\sqrt{\sigma})\ |\Gamma(1-\vartheta+i\sqrt{\sigma})|^2\ |\Gamma(1-\tilde{\vartheta}+i\sqrt{\sigma})|^2.$$

By modifying $W^{(\alpha,\beta)}$ by a factor of norm one we can define the isometry

$$\mathcal{J}^{(\alpha,\beta)}: H_{\alpha,\beta} \longrightarrow L_2((1,\infty)) \tag{2.160}$$

$$\mathcal{J}^{(\alpha,\beta)}(g)(\sigma) :=$$

$$= 2^{-\vartheta-1/2}\ \pi^{-1}\ (\sinh(2\pi\sqrt{\sigma}))^{1/2}\ \Gamma(1-\vartheta+i\sqrt{\sigma})\ \Gamma(1-\tilde{\vartheta}-i\sqrt{\sigma})\ (f_\sigma^{(\alpha,\beta)},g)_{\alpha,\beta}\ ,$$

where $(\bullet,\bullet)_{\alpha,\beta}$ denotes the inner product in $H_{\alpha,\beta}$, $\sigma \in \mathbb{R}^+$, and is here used in a symbolic way (as we did with the Fourier transform) as an abbreviation. Since the kernel function is well-defined for $\alpha \in \mathbb{N} \cup \{0\}$, we may extend $\mathcal{J}^{(\alpha,\beta)}$ to arbitrary $\alpha > -1$. We shall call $\mathcal{J}^{(\alpha,\beta)}$ <u>Jacobi transform</u>.

<u>Theorem 2.7</u>: *The selfadjoint operator* $A_{\alpha,\beta}$, $\alpha > -1$, $|\beta| \le \alpha + 1$, *has a simple spectrum*

$$\sigma(A_{\alpha,\beta}) = C\sigma(A_{\alpha,\beta}) = \overline{\mathbb{R}^+}.$$

The unitary mapping

$$\mathcal{J}^{(\alpha,\beta)}: H_{\alpha,\beta} \longrightarrow L_2(\mathbb{R}^+),$$

is a spectral representation for $A_{\alpha,\beta}$.

Proof: To see that $\mathcal{F}^{(\alpha,\beta)}$ is unitary it remains to show that it is onto. As before we have

$$\mathcal{F}^{(\alpha,\beta)} A_{\alpha,\beta} \Phi = m\, \mathcal{F}^{(\alpha,\beta)} \Phi, \tag{2.161}$$

for $\Phi \in C_{\infty}((1,\infty))$ and so as in the proof of Theorem 2.4 the result follows for $\alpha \geq 1$. Following the idea used in the proof of Theorem 2.6 we find that for $\alpha \in (-1,1)$

$$N(A_{\alpha,\beta} \pm i) = \text{span}\{g_{\mp i}^{(a,\beta)} - \omega(\mp i)\, f_{\mp i}^{(\alpha,\beta)}\}, \tag{2.162}$$

and therefore that

$$D(A_{\alpha,\beta}) = D(A_{\alpha,\beta}) \oplus \text{span}\{\vartheta_{\alpha,\beta}\}, \tag{2.163}$$

with

$$\vartheta_{\alpha,\beta} = g_{+i}^{(a,\beta)} - g_{-i}^{(\alpha,\beta)} - \omega(+i)\, f_{+i}^{(\alpha,\beta)} + \omega(-i)\, f_{-i}^{(\alpha,\beta)},$$

in this case. By partial integration and a density argument we have that (2.154) remains valid for $\Phi \in D(A_{\alpha,\beta})$. For this note that

$$\vartheta_{\alpha,\beta}(x) = O(x^{-\vartheta-1/\sqrt{2}}),\quad \vartheta'_{\alpha,\beta}(x) = O(x^{-\vartheta-1/\sqrt{2}-1}) \text{ as } x \longrightarrow \infty.$$

As in the proof of Theorem 2.4 we obtain that $\mathcal{F}^{(\alpha,\beta)}$ is onto (; the details may be left as an exercise). According to (2.161) for $\Phi \in D(A_{\alpha,\beta})$ we also have that $\mathcal{F}^{(\alpha,\beta)}$ is a spectral representation and that the spectrum is continuous and simple.

□

The inverse of $\mathcal{F}^{(\alpha,\beta)}$ in integral form is easily calculated as

$$\mathcal{F}^{(\alpha,\beta)^*}(g)(x) :=$$

$$= 2^{-\vartheta-1/2}/\pi \int_{\mathbb{R}^+} (\sinh(2\pi\sqrt{\sigma}))^{1/2}\, \Gamma(1-\vartheta-i\sqrt{\sigma})\, \Gamma(1-\tilde{\vartheta}+i\sqrt{\sigma})\, f_{\sigma}^{(\alpha,\beta)}(x)\, g(\sigma)\, d\sigma,$$

for $g \in L_2(\mathbb{R}^+)$.

Remark 2.9: *Choosing $\alpha = \beta = \nu-1/2$ leads to an integral transform with the Gegenbauer function as kernel. The case $\nu = 1/2$, i.e. $\alpha = \beta = 0$, leads to the Mehler transform (; special case of generalized Mehler transform: $\mu = 0$). By introducing the unitary mapping*

$$j: H_{\alpha,\beta} \longrightarrow 2^{-\vartheta} (\sinh t)^{-\alpha-1/2} (\cosh t)^{-\beta-1/2} L_2(\mathbb{R}^+),$$

with

$$j(u)(x) = u(\cosh 2x),\ u \in H_{\alpha,\beta},$$

we see that

$$\mathcal{J}^{(\alpha,\beta)} \circ j^*$$

is a spectral representation of

$$B_{\alpha,\beta} := j\ A_{\alpha,\beta}\ j^*$$

The differential expression $B_{\alpha,\beta}$ *is related to is*

$$\frac{1}{4}(2\pi D)^2 - \left\{(2\alpha+1)\coth(m)/4 + (2\beta+1)\ \tanh(m)\right\}(2\pi iD) - \vartheta^2.$$

From this it is clear that for $\alpha = \beta = -1/2$ *we have that* $\mathcal{J}^{(-1/2,-1/2)} \circ j$ *is a spectral representation related to the differential expression* $(\pi D)^2$*. The boundary condition that has to be satisfied for elements in* $D(A_{-1/2,-1/2})$ *implies*

$$u'(x) \longrightarrow 0 \text{ as } x \longrightarrow 0+,$$

for $u \in D(B_{-1/2,-1/2})$. *Thus we see that* $\mathcal{J}^{(-1/2,-1/2)} \circ j$ *is a spectral representation for the Neumann case for* $(\pi D)^2$ *and therefore must be closely related to the one-sided Fourier cosine transform. In fact we have that the Jacobi function* $\phi_\lambda^{(\alpha,\beta)}$ *defined by*

$$\phi_{2\sqrt{\lambda}}^{(\alpha,\beta)} := j(f_\lambda^{(\alpha,\beta)})\ \Gamma(\alpha+1),\ \lambda \in \mathbb{R}^+,$$

is for $\alpha = \beta = -1/2$ equal to

$$\phi_\lambda^{(\alpha,\beta)}(t) = \cos(\lambda t),$$

(see e.g. [18], p. 5).

Thus, we have

$$(\mathcal{J}^{(-1/2,-1/2)} \circ j)(f)(\sigma) = \chi_0(\sigma)\ F_{\cos}(f)(4\sigma),$$

where

$$\chi_0(\sigma) = 2^{-1/2}\ \pi^{-1}\ \sigma^{1/4}\ \sinh(2\pi\sqrt{\sigma})^{1/2}\ \Gamma(1+i\sqrt{\sigma})\ \Gamma(1/2 - i\sqrt{\sigma}).$$

Noting that the expressions

$$\pi^{-1/2}\ \sigma^{1/4}\ (\sinh(\pi\sqrt{\sigma}))^{1/2}\ \Gamma(i\sqrt{\sigma}),\ \pi^{-1/2}\ (\cosh(\pi\sqrt{\sigma}))^{1/2}\ \Gamma(1/2-i\sqrt{\sigma})$$

have norm one ([23], p. 2) *we see that also*

$$|\chi_0(\sigma)| = 1,\ \sigma \in \mathbb{R}^+.$$

With this example we conclude our investigation of second order differential expressions. It is obvious that the number of possibilities to consider further integral transforms is limitless. Since we have finished our program to give the classically well-known integral transforms their place in our study, we finish our encounter with spectral theory by looking at a fourth order example to demonstrate that the ideas presented carry beyond the limitations imposed so far.

2.7 Excursion: A spectral representation associated with the differential expression $(2\pi D)^4$

We have already studied several integral transforms that can be associated with the differential expression D^4 (Fourier transform, one-sided and two-sided Fourier sine and Fourier cosine transform). Of course there is a whole family of integral transforms if we consider all possible selfadjoint extensions of the operator $\mathcal{A}$ defined as the closure of $(2\pi D)^4$ defined on $C_\infty(\mathbb{R}^+)$. We shall limit our attention to the simple cases of Dirichlet and Neumann boundary conditions (see [20] for the Dirichlet case). Let A be the closure of $(2\pi D)^2$ defined on $C_\infty(\mathbb{R}^+)$ then

$$\mathcal{A}_D := A^* A, \quad \mathcal{A}_N := A A^*, \tag{2.164}$$

The domains $D(\mathcal{A}_D)$, $D(\mathcal{A}_N)$, can —— based on our considerations in section 1.1 —— be characterized very easily. We have

$$D(\mathcal{A}_D) = \{\Phi \in H_1(\mathbb{R}^+) \mid \Phi' \in H_1(\mathbb{R}^+),\ \Phi'' \in H_1(\mathbb{R}^+),\ \Phi''' \in H_1(\mathbb{R}^+)\},$$

$$D(\mathcal{A}_N) = \{\Phi \in H_1(\mathbb{R}^+) \mid \Phi' \in H_1(\mathbb{R}^+),\ \Phi'' \in H_1(\mathbb{R}^+),\ \Phi''' \in H_1(\mathbb{R}^+)\}$$

Possible candidates for elements in $N(\mathcal{A}^* \pm i)$ are

$$e^{-\alpha m},\ e^{\alpha m},\ e^{i\alpha m},\ e^{-i\alpha m},\ \alpha^4 = \pm i.$$

Thus, we have

$$N(\mathcal{A}^* \pm i) = \operatorname{span}\{e^{-\alpha(\pm) m},\ e^{-i\alpha(\pm) m}\},\ \alpha(+) = e^{i\pi/8},\ \alpha(-) = e^{i3\pi/8}. \tag{2.165}$$

The deficiency indices are therefore $(2|2)$. Green's function is a specific solution of

$$((2\pi D)^4 - \lambda)\varphi = (2\pi D - \lambda^{1/4})(2\pi D + \lambda^{1/4})(2\pi D - i\lambda^{1/4})(2\pi D + i\lambda^{1/4})\varphi = \delta. \tag{2.167}$$

We can find a particular solution by using the factorization given in (2.167) and the knowledge that

$$w_\sigma = i\ \mathrm{sgn}(\mathrm{Im}(\sigma))\ (\ell_{\mathrm{Im}(\sigma),0}\ h)\ e^{i\sigma m} \in L_2(\mathbb{R}),\ \sigma \notin \mathbb{R}, \tag{2.168}$$

is a solution of

$$(2\pi D-\sigma)\ w_\sigma = \delta, \tag{2.169}$$

(see 1.1.2; for the notation of $\mathscr{C}_{\alpha,\beta}$ see 1.1.2.1).

Thus we obtain a solution g_λ as

$$g_\lambda = w_{\sigma_1} * w_{-\sigma_1} * w_{\sigma_2} * w_{-\sigma_2}, \tag{2.170}$$

where

$$\sigma_1 = \lambda^{1/4},\ \sigma_2 = i\ \lambda^{1/4},$$

with $\lambda^{1/4} := |\lambda|^{1/4}\ e^{i(\arg\lambda)/4}$, $\lambda \in \mathbb{C}\backslash\mathbb{R}$, $\arg(\lambda) \in [0,2\pi)$.

Evaluation of (2.170) yields

$$g_\lambda(x) = \frac{1}{4\ \sigma^3}\ (-e^{-\sigma|x|} + i\ e^{i\sigma|x|}), \tag{2.171}$$

with $\sigma = \lambda^{1/4}$, $\lambda \notin \mathbb{R}$. Thus we have

$$G_\lambda(x,t) = g_\lambda(x-t) + \frac{1}{4\ \sigma^3}\ (a(t)\ e^{-\sigma x} + b(t)\ e^{i\sigma x}), \tag{2.172}$$

for a suitable choice of $a(t)$, $b(t)$ in order to satisfy the boundary conditions

$$\begin{aligned} &G_{D,\lambda} \in H_1(\mathbb{R}^+),\ G'_{D,\lambda} \in H_1(\mathbb{R}^+), \\ &G''_{N,\lambda} \in H_1(\mathbb{R}^+),\ G'''_{N,\lambda} \in H_1(\mathbb{R}^+), \end{aligned} \tag{2.173}$$

in the Dirichlet and Neumann case respectively. Let us first deal with the Dirichlet boundary condition.

By (2.173) we have

$$-e^{-\sigma t} + i\ e^{i\sigma t} + a(t) + b(t) = 0,$$

$$-e^{-\sigma t} + e^{i\sigma t} - a(t) + i\ b(t) = 0,$$

or

$$a(t) = i\ e^{-\sigma t} + (1-i)\ e^{i\sigma t},$$

$$b(t) = -\ e^{i\sigma t} + (1-i)\ e^{-\sigma t}.$$

In other words

$$G_{D,\lambda}(x,t) = \frac{1}{4\ \sigma^3}\ (-e^{-\sigma|x-t|} + i\ e^{i\sigma|x-t|} + $$

$$+ i\ e^{-\sigma(x+t)} - e^{i\sigma(x+t)} + (1-i)\ e^{i\sigma(x+it)} + (1-i)\ e^{i\sigma(t+ix)}).$$

For $\varepsilon, \lambda \in \mathbb{R}^+$, $\mu = \lambda^{1/4}$, $\sigma_\pm = (\lambda\pm i\varepsilon)^{1/4}$ (in the above specified sense), we have

$$\lim_{\varepsilon \to 0+} \sigma_+ = \mu, \quad \lim_{\varepsilon \to 0+} \sigma_- = i\mu,$$

and so

$$\lim_{\varepsilon \to +0} G_{D,\lambda+i\varepsilon}(x,t) = \frac{1}{4\mu^3}\,(-e^{-\mu|x-t|} + i\, e^{i\mu|x-t|} +$$

$$+ i\, e^{-\mu(x+t)} - e^{i\mu(x+t)} + (1-i)\,(e^{i\mu(x+it)} + e^{i\mu(t+ix)}),$$

and

$$\lim_{\varepsilon \to +0} G_{D,\lambda-i\varepsilon}(x,t) = \frac{1}{4\mu^3}\,(-i\, e^{-i\mu|x-t|} - e^{-\mu|x-t|} +$$

$$- e^{-i\mu(x+t)} - i\, e^{-\mu(x+t)} + (1+i)\,(e^{-\mu(x+it)} + e^{-\mu(t+ix)}).$$

Consequently,

$$\lim_{\varepsilon \to 0} (G_{D,\sigma+i\varepsilon}(x,t) - G_{D,\sigma-i\varepsilon}(x,t)) =$$

$$= \frac{i}{2\mu^3}\,(\cos(\mu(x-t)) - \sin(\mu(x+t)) + e^{-\sigma(x+t)} + \qquad (2.174)$$

$$+ e^{-\mu t}(\sin \mu x - \cos \mu x) + e^{-\mu x}(\sin \mu t - \cos \mu t)),$$

$$= \frac{i}{2\mu^3}\,(e^{-\mu x} + \sin \mu x - \cos \mu x)\,(e^{-\mu t} + \sin \mu t - \cos \mu t).$$

From (2.174) we obtain ($0 < \mu \le \lambda$) for the spectral family $(\Pi_D(\lambda))_{\lambda \in \mathbb{R}}$ of A_D :

$$(\Pi_D(\lambda) - \Pi_D(\mu) f\ ,\ g) = \int_\mu^\lambda \overline{P_D(f)(\sigma)}\, P_D(g)(\sigma)\, d\sigma \qquad (2.175)$$

$f,g \in L_2(\mathbb{R}^+)$, $\operatorname{supp} f \cup \operatorname{supp} g \subset\subset \mathbb{R}^+$,with

$$P_D(f)(\sigma) = \frac{1}{2}\,\sigma^{-3/8} \int_{\mathbb{R}^+} [e^{-\sigma^{1/4}\cdot x} + \sin(\sigma^{1/4}\cdot x) - \cos(\sigma^{1/4}\cdot x)] f(x)\, dx\ .$$

Extension by isometry leads from (2.175) to Parseval's equality

$$(f,g) = (P_D(f), P_D(g))\ ,\ f,g \in L_2(\mathbb{R}^+)\ ,$$

(observe that 0 is not an eigenvalue and $\mathbb{R}^-$ is not in the spectrum).
That

$$P_D : L_2(\mathbb{R}^+) \longrightarrow L_2(\mathbb{R}^+)$$

is not only isometric but also onto follows by similar considerations as in the case of Hankel transform. Thus, we have

Theorem 2.8: *$\mathcal{A}_D$ is a selfadjoint operator with simple spectrum*

$$\sigma(\mathcal{A}_D) = C\sigma(\mathcal{A}_D) = \overline{\mathbb{R}^+} .$$

The integral transform P_D *is a spectral representation of* $\mathcal{A}_D$.

The Neumann case can be dealt with similarly. To satisfy the Neumann boundary condition (see (2.173)) we have

$$-e^{-\sigma t} - i\, e^{i\sigma t} + a(t) - b(t) = 0,$$

$$-e^{-\sigma t} - e^{i\sigma t} - a(t) - i\, b(t) = 0,$$

or

$$a(t) = i\, e^{-\sigma t} - (1-i)\, e^{i\sigma t},$$

$$b(t) = -\, e^{i\sigma t} - (1-i)\, e^{-\sigma t}.$$

In other words

$$G_{N,\lambda}(x,t) = \frac{1}{4\,\sigma^3}\,(-e^{-\sigma|x-t|} + i\, e^{i\sigma|x-t|} +$$
$$+ i\, e^{-\sigma(x+t)} - e^{i\sigma(x+t)} - (1-i)\, e^{i\sigma(x+it)} - (1-i)\, e^{i\sigma(t+ix)}).$$

For $\varepsilon, \lambda \in \mathbb{R}^+$, $\mu = \lambda^{1/4}$, $\sigma_\pm = (\lambda \pm i\varepsilon)^{1/4}$ (in the above specified sense), we have

$$\lim_{\varepsilon \to 0+} \sigma_+ = \mu, \quad \lim_{\varepsilon \to 0+} \sigma_- = i\mu,$$

and so

$$\lim_{\varepsilon \to +0} G_{N,\lambda+i\varepsilon}(x,t) = \frac{1}{4\,\mu^3}\,(-e^{-\mu|x-t|} + i\, e^{i\mu|x-t|} +$$
$$+ i\, e^{-\mu(x+t)} - e^{i\mu(x+t)} - (1-i)\,(e^{i\mu(x+it)} + e^{i\mu(t+ix)}),$$

and

$$\lim_{\varepsilon \to +0} G_{N,\lambda-i\varepsilon}(x,t) = \frac{1}{4\,\mu^3}\,(-i\, e^{-i\mu|x-t|} - e^{-\mu|x-t|} +$$
$$- e^{-i\mu(x+t)} - i\, e^{-\mu(x+t)} - (1+i)\,(e^{-\mu(x+it)} + e^{-\mu(t+ix)}).$$

Consequently,

$$\lim_{\varepsilon \to 0} (G_{\sigma+i\varepsilon}(x,t) - G_{\sigma-i\varepsilon}(x,t)) =$$
$$= \frac{i}{2\,\mu^3}\,(\cos(\mu(x-t)) - \sin(\mu(x+t)) + e^{-\sigma(x+t)} + \tag{2.176}$$
$$+ e^{-\mu t}(\cos \mu x - \sin \mu x) + e^{-\mu x}(\cos \mu t - \sin \mu t)),$$
$$= \frac{i}{2\,\mu^3}\,(e^{-\mu x} + \cos \mu x - \sin \mu x)\,(e^{-\mu t} + \cos \mu t - \sin \mu t).$$

From (2.176) we obtain an analogous result for the Neumann case.

<u>Theorem 2.8</u>: $\mathcal{A}_N$ *is a selfadjoint operator with simple spectrum*

$$\sigma(\mathcal{A}_N) = C\sigma(\mathcal{A}_N) = \overline{\mathbb{R}^+} .$$

The integral transform P_N *defined as the isometric extension of*

$$P_N(f)(\sigma) = \frac{1}{2}\sigma^{-3/8} \int_{\mathbb{R}^+} [e^{-\sigma^{1/4}\cdot x} + \cos(\sigma^{1/4}\cdot x) - \sin(\sigma^{1/4}\cdot x)]\, f(x)\, dx$$

is a spectral representation of $\mathcal{A}_N$.

Part 3: Appendix: Tools from functional analysis

In this appendix we shall give a survey over results and concepts of functional analysis in Hilbert space applied to the theory of integral transforms and eigenfunction expansions in chapter 1 and 2. The results will be given without proof, but appropriate references are provided. The idea of this chapter is to circumscribe the area of functional analysis of which the considerations in the main part of these lecture notes can be regarded as an application of. For this purpose it is enough to restrict our collection of useful results to the Hilbert space aspects of functional analysis. It is not intended to give an introduction into functional analysis but to provide a reminder of functional analytical concepts. General references for the appendix are [2], [6], [10], [11], [17], [30], [33], [37].

A (complex) Hilbert space is a linear space (over the complex numbers $\mathbb{C}$) with inner product, complete with respect to the induced norm. As a convention we assume that the inner product will always be linear in the second factor. A standard construction is by completion of a pre-Hilbert space (a not necessarily complete, linear space with inner product). There are two other standard construction principles of importance to build Hilbert spaces from (pre-) Hilbert spaces: the direct sum and the tensor product. Let $(H_\alpha)_{\alpha\in A}$ be a family of (pre-) Hilbert spaces, A an index set, then we consider the cartesian product $\Pi = \prod_{\alpha\in A} H_\alpha$, i.e. the set of all mappings from A to $\bigcup_{\alpha\in A} H_\alpha$ such that

$$f(\alpha) \in H_\alpha .$$

$f \in \Pi$ may be identified with a function $\delta_f\colon \Pi \to \mathbb{C}$, $\delta_f(f) = 1$, $\delta_f(h) = 0$ for $h \in \Pi\backslash\{f\}$.

The set

$$\Sigma_0 = \{\delta_f \mid f \in \Pi \ \&\ \{\alpha \in A \mid f(\alpha) \neq 0\} \text{ finite}\} \tag{A.1}$$

regarded as a subset of complex-valued functions from Π into $\mathbb{C}$ has a natural linear structure, so that the linear span of Σ_0 in the set of complex-valued functions (as a linear space) may be used to define a linear space

$$\sum_{\alpha\in A} H_\alpha := \text{span } \Sigma_0 ,$$

the weak direct sum of $(H_\alpha)_{\alpha\in A}$. The weak direct sum $\sum_{\alpha\in A} H_\alpha$ can be equipped with a semi-norm $\|\bullet\|_\oplus = ((\bullet,\bullet)_\oplus)^{1/2}$ induced by the hermitian sesqui-linear form generated by the definition

$$(f,g)_\oplus := \sum_{\alpha\in A} (f(\alpha),g(\alpha))_{H_\alpha} , \quad f,g \in \Sigma_0 , \tag{A.2}$$

here $(\bullet,\bullet)_{H_\alpha}$ denotes the inner product in H_α (observe that the r.h.s. of equation (A.2) is always finite). Recall that a bounded hermitian sesqui-linear form on a Hilbert space H is a mapping from $H \times H$ into $\mathbb{C}$ such that $\beta(x,\bullet)$ is linear for every $x \in H$, $\beta(x,y) = \overline{\beta(y,x)}$ for all $x,y \in H$ and $|\beta(x,y)| \leq c \ \|x\| \ \|y\|$ for all $x,y \in H$, ($\|\bullet\|$ denotes the norm in H).
The set

$$N_\oplus := \{f \in \sum_{\alpha\in A} H_\alpha \mid \|f\|_\oplus = 0\}$$

is a linear subspace and we may thus consider the <u>algebraic direct sum</u>

$$\sum_{\alpha\in A} H_\alpha \Big/ N_\oplus$$

as a pre-Hilbert space with the inner product induced by (A.2) . If $f \in \Sigma_0$ then the corresponding element in the quotient space is often denoted by

$$\bigoplus_{\alpha\in A} f(\alpha) .$$

The completion of the pre-Hilbert space $\left(\sum_{\alpha\in A} H_\alpha \Big/ N_\oplus \ \middle| \ (\bullet,\bullet)_\oplus \right)$ is called the direct sum $\bigoplus_{\alpha\in A} H_\alpha$ of the family of Hilbert spaces $(H_\alpha)_{\alpha\in A}$.

In the sense of this definition we have e.g. $\mathbb{C}^n = \bigoplus_{i=1}^{n} \mathbb{C}$, if we consider $\mathbb{C}$ as a Hilbert space with inner product $\overline{a}\bullet b$, $a,b \in \mathbb{C}$. The linear structure of $\bigoplus_{\alpha\in A} H_\alpha$ coincides with the natural linear structure of the elements of Π as point-wise members of a linear space

$$(f+a\ g)(\alpha) = f(\alpha) + a\ g(\alpha) , \quad \alpha \in A , \ a \in \mathbb{C}., \tag{A.3}$$

$f,g \in \Pi$. In this sense the elements of $\bigoplus_{\alpha\in A} H_\alpha$ can be identified with

$$\{f \in \Pi \mid \{\alpha \in A \mid f(\alpha) \neq 0\} \text{ countable } \& \ \|f\|_{\oplus} < \infty\}$$

as a linear space in the sense of (A.3) .

The direct product or tensor product is another construction based on the weak direct sum. The linear space $\sum_{\alpha\in A} H_\alpha$ may be equipped with a semi-norm $\|\bullet\|_{\otimes}$ induced by the hermitian sesqui-linear form defined by

$$(f,g)_{\otimes} = \prod_{\alpha\in A} (f(\alpha),g(\alpha))_{H_\alpha} , \quad f,g \in \Sigma_0 . \tag{A.4}$$

The set

$$N_{\otimes} = \{f \in \sum_{\alpha\in A} H_\alpha \mid \|f\|_{\otimes} = 0\}$$

proves to be a linear subspace. Thus, the algebraic tensor product

$$\sum_{\alpha\in A} H_\alpha \Big/ N_{\otimes}$$

is a pre-Hilbert space with respect to the inner product induced by (A.4) . The element of this quotient corresponding to an element $f \in \Sigma_0$ is written as

$$\bigotimes_{\alpha\in A} f(\alpha) .$$

The completion of this pre-Hilbert space is called the tensor product $\bigotimes_{\alpha\in A} H_\alpha$ of $(H_\alpha)_{\alpha\in A}$. It should be noted that the concept of tensor product is only useful if A is finite ($\bigotimes_{\alpha\in A} H_\alpha = \{0\}$ for non-finite A). To illustrate this definition we note that the linear space of $(n|n)$-matrices can be regarded as $\mathbb{C}^n \otimes \mathbb{C}^n$, and that the space $L_2(\mathbb{R}^n)$ of measurable, square integrable, complex-valued functions over $\mathbb{R}^n$ can be identified with

$$\bigotimes_{i\in\{1,2,\dots,n\}} L_2(\mathbb{R}) .$$

We note the following fact about fundamental sets (i.e. sets with dense span) in the direct sum and tensor product space respectively.

<u>Theorem A.1</u>: *Let* M_α *be fundamental in* H_α , $\alpha \in A$, *A index set, then*

$$\{ \bigoplus_{\alpha\in A} f(\alpha) \mid f(\alpha) \in M_\alpha , \alpha \in A , f \in \Sigma_0\}$$

is fundamental in $\bigoplus_{\alpha\in A} H_\alpha$

and

$$\{ \bigotimes_{\alpha\in A} f(\alpha) \mid f(\alpha) \in M_\alpha , \alpha \in A , f \in \Sigma_0 \}$$

is fundamental in $\bigotimes_{\alpha\in A} H_\alpha$.

([37], p. 52-53).

As a central theorem of Hilbert space theory we have the projection theorem.

Theorem A.2: *Let* M *be a closed subspace of a Hilbert space* H *with inner product* $(\cdot,\cdot)$ *and* $M^\perp = \{x \in H \mid (x,y) = 0$ for all $y \in M\}$, *then every element* $f \in H$ *can be written as*

$$f = u + v , u \in M , v \in M^\perp ; \tag{A.5}$$

$u,v \in H$ *are uniquely determined.*

This result is sometimes written as

Corollary A.2: *Let* M *be a closed subspace of a Hilbert space* H , *then*

$$H = M \oplus M^\perp . \tag{A.6}$$

The equality sign in (A.6) has to be understood as unitary equivalence, i.e. there is a norm-preserving mapping $j: H \longrightarrow M \oplus M^\perp$ onto $M \oplus M^\perp$ (observe that $M,M^\perp$ are Hilbert spaces). The theorem above gives rise to the definition of an operator

$$P : H \longrightarrow H$$

mapping any element $f \in H$ onto the element $u \in M$ according to (A.5) . This operator is called an orthogonal projection on M and satisfies

$$\begin{aligned} &P P = P , \\ &\|Px\| \le \|x\| , \\ &(Px,y) = (x,Py) , x,y \in H . \end{aligned} \tag{A.7}$$

An important consequence of the projection theorem is the Riesz representation theorem.

Theorem A.3: *Let the linear mapping* $F: H \longrightarrow \mathbb{C}$, H *Hilbert space, be bounded, then there is a unique* $f \in H$ *such that*

$$F(x) = (f,x) .$$

The mapping $F \longrightarrow f$ *is called Riesz map and is a norm-preserving mapping in the sense that*

$$\|f\| = \sup_{\substack{x \in H \\ x \neq 0}} \frac{|F(x)|}{\|x\|}.$$

Let now $A: D(A) \subset H \longrightarrow H$ be a linear operator from the linear subspace $D(A)$ of the Hilbert space H into H . If A is densely defined (i.e. $\overline{D(A)} = H$) then we can define its adjoint A^* by applying Theorem A.3.

The linear functional F_x

$$F_x(y) := (Ay,x) \ , \ y \in D(A) \ ,$$

may be bounded with respect to $y \in D(A)$ such that

$$|F_x(y)| \leq C \ \|y\| \ ,$$

for some constant $C > 0$. Let $\mathbf{F}_x$ denote its closure, i.e.

$$\mathbf{F}_x(y) := \lim_{n\to\infty} F_x(y_n) \ , \ y_n \longrightarrow y \text{ in } H \text{ as } n \to \infty \ ,$$

then we say: $x \in D(A^*)$ iff $\mathbf{F}_x$ is bounded linear mapping from H into $\mathbb{C}$.

Theorem A.3 leads to the characterization

$$D(A^*) = \{x \in H \mid (Ay,x)=(y,f) \text{ for all } y \in D(A) \text{ and some } f \in H\}. \qquad (A.8)$$

The element f in this characterization is given by the Riesz map R applied to $\mathbf{F}_x$. Now we define the adjoint of A by

$$A^*x = R\ \mathbf{F}_x \text{ for } x \in D(A^*) \ . \qquad (A.9)$$

Theorem A.3 also provides the means to assign an operator to densely defined, semi-bounded, hermitian sesqui-linear forms. A hermitian sesqui-linear form β is called semi-bounded (from below) if

$$\beta(x,x) \geq c \ \|x\|^2$$

for some $c \in \mathbb{R}$ and all x in the domain of definition $D(\beta) \subset H$ of β , $\|\cdot\|$ denotes the norm in the Hilbert space H with inner product $(\cdot,\cdot)$.

The sesquilinear form

$$\beta_0(x,y) = \beta(x,y) + (1-c)(x,y)$$

is now an inner product for $D(\beta)$ such that $(D(\beta)|\beta_0)$ can be regarded as a pre-Hilbert space. Let H_β denote its completion. Since

$$\beta_0(x,x) \geq (x,x) ,$$

for all $x \in D(\beta)$, H_β may be regarded as subspace of H . If β can be extended to a sesquilinear form $\overline{\beta}$ on H_β by the definition

$$\overline{\beta}(x,x) := \lim_{n\to\infty} \beta(x_n,x_n) ,$$

for $x_n \longrightarrow x$ in H as $n \longrightarrow \infty$, where $(x_n)_n \subset D(\beta)$ is a Cauchy sequence in H_β , then β is called closable and $\overline{\beta}$ is its closure. In this case we have $D(\overline{\beta}) = H_\beta$. The term <u>symmetric</u> sesqui-linear form (operator) is used for a densely defined, hermitian sesqui-linear form (operator). A linear operator A is called hermitian if

$$(Ax,y) = (x,Ay) \text{ for all } x,y \in D(A) .$$

If a symmetric operator A satisfies $D(A) = D(A^*)$, then we say A is <u>selfadjoint</u>. We have the following connection between symmetric sesquilinear forms and selfadjoint operators:

<u>Theorem A.4</u>: *Let* β *be a closable, symmetric, semi-bounded, sesquilinear form then there is a unique selfadjoint operator* A *with*

$$D(A) = \{x \in H_\beta \mid \overline{\beta}(g,x) = (g,f) \text{ for all } g \in D(\beta) \text{ and some } f \in H\}$$

such that

$$(y,Ax) = \overline{\beta}(y,x) .$$

The operator A *is semi-bounded with the same constant as* β .

On the other hand we have for a semibounded operator, i.e.

$$(x,Ax) \geq c \, \|x\|^2 , \; x \in D(A) , \text{ for a constant } c \in \mathbb{R} ,$$

the following theorem.

<u>Theorem A.5</u>: *Let* A *be a symmetric, semi-bounded operator then*

$$\beta(x,y) := (x,Ay) , \; x,y \in D(A) \tag{A.10}$$

is a semi-bounded, symmetric sesquilinear form. Furthermore, β *is closable.*

The discussion of selfadjoint extension of symmetric operators is based on a relation between symmetric and norm-preserving operators. The mapping C from the set of symmetric operators into the set of norm-preserving linear operators (isometries) defined by

$$C(A) = (A+i)(A-i)^{-1}$$

is called Cayley transform. C is one-to-one and onto. The inverse is given by

$$C^{-1}(I) = -i\,(I+1)(I-1)^{-1}\ ,\ I \text{ isometry}.$$

Before we formulate the properties of C in detail we have to recall the concept of closability for linear operators. The closability of a linear operator can be seen in analogy to the closability of sesqui-linear forms. The domain $D(A)$ of a linear operator $A: D(A) \subset H_1 \longrightarrow H_2$, H_i Hilbert space $i=1,2$, can be regarded as a pre-Hilbert space with respect to

$$(x,y) + (Ax,Ay)$$

as inner product. Let H_A denote its completion; H_A can be imbedded in H_1 . If A can be extended to a linear operator A defined on H_A by the definition

$$\bar{A}x := \lim_{n\to\infty} Ax_n \text{ (convergence in } H_2 \text{)} ,$$

for any Cauchy sequence $(x_n)_n \subset D(A)$ in H_A such that $x_n \longrightarrow x$ in H_1 as $n \longrightarrow \infty$, then A is called closable, $\bar{A}$ is its closure. A is closed if $A = \bar{A}$. We note the following results

<u>Theorem A.6</u>: *Let* $A: D(A) \subset H \to H$ *be a linear densely defined operator in a Hilbert space* H *then*

$$A \textit{ closable iff } A^* \textit{ densely defined,}$$

$$A^* \textit{ is closed.}$$

If A *is closable, we have*

$$H = N(\bar{A}) \oplus \overline{W(A^*)} = N(A^*) \oplus \overline{W(A)}$$

and

$$\bar{A} = A^{**} .$$

The theorem about Cayley's transform will be stated next.

<u>Theorem A.7</u>: *If* A *is a hermitian operator, then* $C(A)$ *is an isometry of* $R(A-i)$ *onto* $R(A+i)$, $R(1-C(A)) = D(A)$. *If* I *is an isometry from* $D(I)$ *onto* $R(I)$ *then* $C^{-1}(I)$ *is hermitian. In particular, if* A *is symmetric then* $R(1-C(A))$ *is dense*

in H ; *if* R(I) *is dense in* H *then* $C^{-1}(I)$ *is symmetric. Furthermore, if* B *is a symmetric continuation of* A *(i.e.* B *symmetric,* $D(A) \subset D(B)$ *and* $Ax = Bx$ *on* D(A)*) then* C(B) *is an isometric continuation of* C(A) .

The following assertions are equivalent:

a) A *closed,*

b) C(A) *closed,*

c) R(A+i) *closed,*

d) R(A-i) *closed.*

The operator A *is selfadjoint if* $C(A)C(A)^*$ *is the identity on* H *(i.e.* C(A) *is unitary).*

Conversely, let I *be unitary then* $C^{-1}(I)$ *is selfadjoint. The following assertions are equivalent:*

a') A *selfadjoint,*

b') C(A) *unitary,*

c') R(A+i) = R(A-i) = H ,

d') $\dim N(A^*-i) = \dim N(A^*+i) = 0$.

If A *is symmetric and* U *is an isometric continuation of* C(A) *then* $B = C^{-1}(U)$ *is a symmetric continuation of* A . *In particular,*

$$D(B) = D(A) \oplus (1-U)N(A^*+i).$$

([37], p. 218-221).

The numbers

$$\alpha_{\pm} := \dim N(A^* \pm i)$$

are called deficiency indices. The quantity $\alpha(\lambda) = \dim N(A^*-\lambda)$ is constant on each connected component of the complement of the closure of the numerical range of A^* (i.e. on $\{t \in \mathbb{C} \mid (A^*x,x) = t,\ x \in D(A^*)\}$).

<u>Corollary A.7</u>: *A symmetric operator* A *admits a selfadjoint extension iff*

$$\alpha_+ = \alpha_- .$$

The most important property of a selfadjoint operator is that it permits a spectral resolution. A family of orthogonal projections $(\Pi(\lambda))_{\lambda \in \mathbb{R}}$ is called spectral family if

1) $\Pi(\mu)\Pi(\lambda) = \Pi(\mu)$, if $\lambda \leq \mu$ (monotony),

2) $\Pi(\lambda+)\Phi = \Pi(\lambda)\Phi$, $\lambda \in \mathbb{R}$, $\Phi \in H$, (strong right continuity),

3) $\Pi(+\infty)\Phi = \Phi$, $\Phi \in H$,

4) $\Pi(-\infty)\Phi = 0$, $\Phi \in H$.

The family $(\Pi(\lambda))_{\lambda\in\mathbb{R}}$ is called a spectral resolution for a selfadjoint operator if

$$(Ax,x) = \int_{\mathbb{R}} \lambda \, d(\Pi(\lambda)x,x) \quad , \quad x \in D(A) \; ,$$

in the sense of a Lebesgue-Stieltjes integral.

Theorem A.8: *Every selfadjoint operator* A *admits a spectral resolution* $(\Pi(\lambda))_{\lambda\in\mathbb{R}}$. *The domain* D(A) *can be characterized by*

$$D(A) = \{x \in H \mid \int_{\mathbb{R}} \lambda^2 \, d\|\Pi(\lambda)x\|^2 < \infty\} \; .$$

The spectral family is related to A *by the formula (Stone's formula)*

$$([\frac{\Pi(\lambda+) + \Pi(\lambda-)}{2} - \frac{\Pi(\mu+) + \Pi(\mu-)}{2}]x \, , \, y) \tag{A.11}$$

$$= \lim_{\varepsilon\to 0+} \int_{\mu}^{\lambda} (\frac{1}{2\pi i} [(A-(\sigma+i\varepsilon))^{-1} - (A-(\sigma-i\varepsilon))^{-1}]x \, , \, y) \, d\sigma.$$

([11], p. 1192, p. 1203, [37], p. 181).

For a linear operator $A : D(A) \subset H \longrightarrow H$ the question of solvability of the equation

$$(A-\lambda)u = f \quad , \quad u \in D(A) \quad , \quad f \in H \quad , \quad \lambda \in \mathbb{C} \; , \tag{A.12}$$

gives rise to a classification of points in $\mathbb{C}$.

The set $\rho(A)$ where reasonable solutions of (A.12) exist is called resolvent set:

$$\rho(A) := \{\lambda \in \mathbb{C} \mid \text{The resolvent } (A-\lambda)^{-1} \text{ exists as a bounded, densely defined operator}\}.$$

With the abreviations

$\alpha(A) :\Leftrightarrow (A-\lambda)^{-1}$ exists,

$\beta(A) :\Leftrightarrow (A-\lambda)^{-1}$ is densely defined,

$$\gamma(A) :\Leftrightarrow (A-\lambda)^{-1} \text{ bounded},$$

we have

$$\rho(A) = \{\lambda \in \mathbb{C} \mid \alpha(A) \;\&\; \beta(A) \;\&\; \gamma(A)\}.$$

$\rho(A)$ is an open set. The spectrum $\sigma(A) := \mathbb{C}\backslash\rho(A)$ is divided into several parts:

the point spectrum $P\sigma(A) := \{\lambda \in \mathbb{C} \mid \text{not } \alpha(A)\}$,

the residual spectrum $R\sigma(A) := \{\lambda \in \mathbb{C} \mid \alpha(A) \;\&\; \text{not } \beta(A)\}$,

the continuous spectrum $C\sigma(A) := \{\lambda \in \mathbb{C} \mid \alpha(A) \;\&\; \beta(A) \;\&\; \text{not } \gamma(A)\}$,

([6], p 298).

The isolated points λ of the spectrum $\sigma(A)$ with $\dim N(A-\lambda) < \infty$ are called discrete spectrum $\sigma_d(A)$. In the study of selfadjoint operators the rest of the spectrum is called essential spectrum

$$\sigma_e(A) := \sigma(A)\backslash\sigma_d(A),$$

([37], p. 191).

For selfadjoint operators we have the situation described in the following theorem.

<u>Theorem A.9</u>: *If A is a selfadjoint operator the spectrum $\sigma(A)$ is real:*

$$\sigma(A) \subset \mathbb{R},$$

$$\mathbb{C}\backslash\mathbb{R} \subset \rho(A),$$

$$R\sigma(A) = \emptyset.$$

<u>Theorem A.10</u>: *Let A be selfadjoint and $(A-\lambda)^{-1}$ be a compact operator for some $\lambda \in \mathbb{C}$, then*

$$\sigma(A) = P\sigma(A) = \sigma_d(A)$$

or

$$\sigma_e(A) = \emptyset, \quad C\sigma(A) = \emptyset.$$

In this case $(A-\lambda)^{-1}$ is compact for every $\lambda \in \rho(A)$.

The spectrum can be characterized by properties of the spectral family.

<u>Theorem A.11</u>: *Let A be a selfadjoint operator, $(\Pi(\lambda))_{\lambda\in\mathbb{R}}$ its spectral resolution then*

$\lambda \in \rho(A) \cap \mathbb{R}$ *iff* $\Pi(\cdot)$ *is constant in a neighbourhood of* $\lambda \in \mathbb{R}$,

$\lambda \in P\sigma(A)$ *iff* $\Pi(\lambda+) - \Pi(\lambda-) \neq 0$; *in this case* $\Pi(\lambda+) - \Pi(\lambda-)$ *is the projection on the eigenspace* $N(A-\lambda)$,

$\lambda \in C\sigma(A)$ *iff* $\Pi(\cdot)$ *is continuous and not constant in a neighbourhood of* $\lambda \in \mathbb{R}$.

The spectral resolution can be used to define functions of an operator:

$$(x, f(A)x) := \int_{\mathbb{R}} f(\lambda)\, d\|\Pi(\lambda)x\|^2 , \qquad \text{(A.13)}$$

for a complex Borel function f defined μ_x-almost everywhere, (for every $x \in H$, $\mu_x(\lambda) := \|\Pi(\lambda)x\|^2$, $\lambda \in \mathbb{R}$) on $\mathbb{R}$ and all $x \in D(f(A))$.

$$D(f(A)) = \{x \in H \mid \int_{\mathbb{R}} |f(\lambda)|^2\, d\|\Pi(\lambda)x\|^2 < \infty\} .$$

The linear operator f(A) is uniquely determined by relation (A.13) (polarization, Riesz representation theorem).

<u>Theorem A.12</u>: *Let* A *be a selfadjoint operator on a Hilbert space* H , $(\Pi(\lambda))_{\lambda\in\mathbb{R}}$ *its spectral resolution, then*

(a) f(A) *closed, densely defined, linear,*

(b) $f(A)^* = \bar{f}(A)$, $(\bar{f}(\lambda) = \overline{f(\lambda)}$, $\lambda \in \mathbb{R}$)

for any complex-valued Borel function f *defined on* $\mathbb{R}$. *In particular,*

$$(x, (A-\lambda)^{-1}x) = \int_{\mathbb{R}} \frac{1}{\mu-\lambda}\, d\|\Pi(\mu)x\|^2 , \quad \lambda \in \rho(A) , \quad x \in H ,$$

$$(x, A^n x) = \int_{\mathbb{R}} \lambda^n\, d\|\Pi(\lambda)x\|^2 , \quad x \in D(A^n) , \quad n \in \mathbb{N} ,$$

the spectral resolution of A^2 *is given by*

$$\Pi(\lambda^{1/2}) - \Pi(-\lambda^{1/2} - 0) , \quad \lambda \in \mathbb{R}^+ .$$

([11], p. 1196-1200, [2], p. 300).

Theorem A.13: *Let* A *be a selfadjoint operator in a Hilbert space* H , $(\Pi(\lambda))_{\lambda\in\mathbb{R}}$ *its spectral resolution and* f,g *complex-valued Borel functions defined on* $\mathbb{R}$, *then*

(a) $(\alpha f)(A) = \alpha f(A)$,

(b) $(f+g)(A)x = f(A)x + g(A)x$ *for* $x \in D(f(A)) \cap D(g(A))$,

(c) $(f\bullet g)(A)x = f(A)\ g(A)x$ *for* $x \in D(f(A)\ g(A)) = D((f\bullet g)(A)) \cap D(g(A))$,

(d) $f(A)\ \Pi(\lambda)x = \Pi(\lambda)f(A)x$ *for* $x \in D(f(A))$.

([11], p. 1199).

A consequence of the spectral theorem, Theorem A.8, is the existence of a spectral representation for any selfadjoint operator. A spectral representation of a selfadjoint operator is a unitary mapping V from the underlying Hilbert space H onto a space of the type

$$\bigoplus_{\alpha\in E} L_2(\mathbb{R},\mu_\alpha)$$

for some index set E , ($L_2(\mathbb{R},\mu_\alpha)$ denotes the space of measurable complex valued functions over $\mathbb{R}$ square integrable with respect to the measure μ_α), satisfying

$$V(A\Phi)(\lambda) = \lambda\ V(\Phi)(\lambda)\ ,\ \lambda \in \mathbb{R}\ . \qquad (A.14)$$

The equality in (A.14) has to be understood in the sense of $\bigoplus_{\alpha\in E} L_2(\mathbb{R},\mu_\alpha)$.

We have the theorem:

Theorem A.14: *Every selfadjoint operator* A *admits a spectral representation* V . *For every Borel function* f *defined on* $\mathbb{R}$ *we have*

$$V(F(A)\Phi)(\lambda) = F(\lambda)\ V(\Phi)(\lambda)\ ,\ \lambda \in \mathbb{R} \qquad (A.15)$$

for

$$\Phi \in D(F(A)) = \{x \in H \mid \sum_{\alpha\in E} \int_{\mathbb{R}} |F(\lambda)\ V_\alpha(\Phi)(\lambda)|^2\ d\mu_\alpha < \infty\}\ ,$$

where V_α *is the projection of* V *onto* $L_2(\mathbb{R},\mu_\alpha)$.

([37], p. 179 et passim, [11], p. 1208 et passim, compare [30], p. 231 et passim).

Corollary A.14: *Let* A *be a selfadjoint operator in* H *and* V *be its spectral representation, then*

$$V((A-z)^{-1}\Phi)(\lambda) = \frac{1}{\lambda-z} V(\Phi)(\lambda) \quad , \quad \lambda \in \mathbb{R} \; , \; z \in \rho(A) \; , \tag{A.16}$$

$$V((\Pi(\sigma) - \Pi(\mu))\Phi)(\lambda) = \chi_{(\mu,\sigma]}(\lambda) \; V(\Phi)(\lambda) \quad , \quad \lambda \in \mathbb{R} \; , \; \mu < \sigma \; , \tag{A.17}$$

where χ_I *denotes the characteristic function of the interval* I .

(Compare [39], p. 118, for an application).

We note that (A.15) can be concluded from (A.17) by an approximation procedure. The connection between (A.16) and (A.17) is given by Stone's formula (A.11) . Equality (A.16) follows directly from (A.14) . In close connection with spectral representations we have the concept of multiplicity. If the index set E is finite, then A (or $\sigma(A)$) is said to be of finite multiplicity. A is said to be of uniform multiplicity if E is finite and all μ_α can be chosen to be equal. In this case the cardinality of E is uniquely determined and named the multiplicity of A (or of its spectrum $\sigma(A)$). If E consists only of one element then we say the spectrum of A is simple.

It is possible to give spectral theorems not only for a single selfadjoint operator but also for a family of commuting selfadjoint operators. In general two closed, linear operators A,B are said to commute if their resolvents commute

$$(A-\lambda)^{-1}(B-\mu)^{-1} = (B-\mu)^{-1}(A-\lambda)^{-1} \quad , \quad \lambda \in \rho(A) \quad , \quad \mu \in \rho(B) \; .$$

For selfadjoint operators we have

Theorem A.15: *If* A,B *are selfadjoint operators on a Hilbert space* H , *then the following statesments are equivalent:*

(a) A,B *commute,*

(b) *the spectral resolution* $(\Pi_A(\lambda))_{\lambda\in\mathbb{R}}$, $(\Pi_B(\lambda))_{\lambda\in\mathbb{R}}$ *of* A,B *resp. commute*

$$\Pi_A(\lambda) \; \Pi_B(\mu) = \Pi_B(\mu) \; \Pi_A(\lambda) \quad , \quad \lambda,\mu \in \mathbb{R} \; ,$$

(c) *every bounded function of* A *commutes with every bounded function of* B.

([30], p. 271-274).

A family $(A_\alpha)_{\alpha\in\Theta}$ of selfadjoint operators is called an <u>abelian system</u> if A_α commute with A_β for any α, β in the index set ϑ. In the following we assume Θ to be finite say $\#\Theta = n$.

The idea of a spectral family can be generalized to more than one spectral parameter.

For $\lambda = (\lambda_1|\lambda_2|\ldots|\lambda_n)$, $\mu = (\mu_1|\mu_2|\ldots|\mu_n) \in \mathbb{R}^n$ we define

$$\lambda \leq \mu \text{ if } (-\infty,\lambda_1]\times\ldots\times(-\infty,\lambda_n] \subset (-\infty,\mu_1]\times\ldots\times(-\infty,\mu_n] . \tag{A.18}$$

An n-parameter spectral family $(\Pi(\lambda))_{\lambda\in\mathbb{R}^n}$ is a family of orthogonal projections satisfying

(1) $E(\mu)\ E(\lambda) = E(\mu)$ for $\mu \leq \lambda$ (monotony),

(2) $E(\lambda+)x = E(\lambda)x$, $x \in H$, $\lambda \in \mathbb{R}^n$ (strong continuity from above) ,

(3) $E(\infty)x = x$, $x \in H$,

(4) $E(-\infty)x = 0$, $x \in H$,

where the limits have to be understood in the sense of (A.18) .

([28], p. 231 et passim).

<u>Theorem A.16</u>: *For every abelian system of selfadjoint operators* $(A_\alpha)_{\alpha=1,\ldots,n}$ *we have an n-parameter spectral family* $(\Pi(\lambda))_{\lambda\in\mathbb{R}^n}$ *such that*

$$(x,A_\alpha x) = \int_{\mathbb{R}^n} \lambda_\alpha \, d\|\Pi(\lambda)x\|^2 , \quad \alpha = 1,2\ldots,n ,$$

for $x \in D(A_\alpha)$

$$D(A_\alpha) = \{x \in H \mid \int_{\mathbb{R}^n} |\lambda_\alpha|^2 \, d\|\Pi(\lambda)x\|^2 < \infty\} .$$

The spectral resolution $(\Pi(\lambda))_{\lambda\in\mathbb{R}^n}$ *of* $(A_\alpha)_{\alpha=1,\ldots,n}$ *can be expressed in terms of the spectral resolutions* $(\Pi_\alpha(\lambda))_{\lambda\in\mathbb{R}}$ *of the* A_α , $\alpha = 1,2\ldots,n$:

$$\Pi(\lambda) = \Pi(\lambda_1)\ldots\Pi(\lambda_n) , \quad \lambda = (\lambda_1|\ldots|\lambda_n) \in \mathbb{R}^n .$$

([28], p. 270 et passim, [33], p. 46-48, p. 60 et passim).

Analogous to the one parameter case we may define functions of an abelian system of selfadjoint operators.

Theorem A.17: *Let* $A = (A_i)_{i=1,\dots,n}$ *be an abelian system of selfadjoint operators on a Hilbert space* H , $(\Pi(\lambda))_{\lambda\in\mathbb{R}^n}$ *its spectral resolution and* F *a complex-valued Borel function on* $\mathbb{R}^n$, *then* F(A) *defined by*

$$(x,F(A)x) = \int_{\mathbb{R}^n} F(\lambda)\, d\|\Pi(\lambda)x\|^2 ,$$

for $x \in D(F(A))$,

$$D(F(A)) = \{x \in H \mid \int_{\mathbb{R}^n} |F(\lambda)|^2\, d\|\Pi(\lambda)x\|^2 < \infty\} ,$$

is a closed, densely defined operator. F(A) *satisfies*

$$F(A)^* = \bar{F}(A) ,$$

such that if F *is real valued,* $B := F(A)$ *is selfadjoint. In this case the spectral resolution* $(\Pi_B(\lambda))_{\lambda\in\mathbb{R}}$ *of* B *is determined by*

$$(x,\Pi_B(\lambda)x) = \int_{\substack{F(\mu)\le\lambda \\ \mu\in\mathbb{R}^n}} d\|\Pi(\mu)x\|^2 , \quad x \in H .$$

([33], p. 60 et passim, [28], p. 270 et passim).

Theorem A.18: *Let* F_i , i=1,2 , *be any complex-valued Borel functions defined on* $\mathbb{R}^n$ *and* $A = (A_i)_{i=1,\dots,n}$ *be an abelian system of selfadjoint operators, then*

$$F_1(A)x + F_2(A)x = (F_1+F_2)(A)x \text{ for } x \in D(F_1(A)) \cap D(F_2(A)) ,$$

$$F_1(A)\, F_2(A)x = (F_1\, F_2)(A)x \text{ for } x \in D(F_1(A)\, F_2(A)) .$$

For any Borel function F *on* $\mathbb{R}^n$ *we have*

$$F(A)^m = (F^m)(A) , \quad m \in N .$$

([28], p. 277 et passim).

Theorem A.19: *Let* $F : \mathbb{R}^n \to \mathbb{R}^m$, $G : \mathbb{R}^m \to \mathbb{R}$, $m,n \in \mathbb{N}$, *be Borel functions then we have for an abelian system* $A = (A_i)_{i=1,\dots,n}$ *with spectral resolution* $(\Pi(\lambda))_{\lambda\in\mathbb{R}^n}$

(a) $F(A) = (F_i(A))_{i=1,\dots,m}$ *is an abelian system,*

and

(b) $G(F(A)) = (G\circ F)(A)$.

The spectral resolution $(\Pi_B(\lambda))_{\lambda\in\mathbb{R}^n}$ *of the abelian system* $B = F(A)$ *is given by*

$$(x,\Pi_B(\lambda)x) = \int_{F(\mu)\leq\lambda} d\|\Pi(\lambda)x\|^2 .$$

([33], p. 61-62).

As a direct application of the 2-parameter case we obtain a spectral theorem for normal operators. A closed, densely defined linear operator A on a Hilbert space H is called normal if $D(A) = D(A^*)$ and

$$AA^* = A^*A .$$

Theorem A.20: *Let* A *be a normal operator on* H , *then*

$$\mathrm{Re}\,A := \frac{1}{2}\,\overline{(A+A^*)}\ ,\quad \mathrm{Im}\,A := \frac{1}{2i}\,\overline{(A-A^*)}$$

are selfadjoint and commuting. For the operator A *we obtain the spectral representation*

$$(x,Ax) = \int_{\mathbb{C}} \lambda\, d\|\Pi(\lambda)x\|^2 ,\quad x\in D(A) ,$$

where

$$\Pi(\lambda+i\mu) := E(\lambda,\mu) ,\quad \lambda,\mu\in\mathbb{R}$$

with $(E(\sigma))_{\sigma\in\mathbb{R}^2}$ *denoting the spectral resolution of the abelian system* $(\mathrm{Re}\,A|\mathrm{Im}\,A)$ *. It is*

$$D(A) = D(\mathrm{Re}\,A) \cap D(\mathrm{Im}\,A) .$$

([37], p. 202).

A standard procedure for constructing commuting sets of selfadjoint operators is by using tensor products.

Theorem A.21: *Let* A_1, A_2 *be selfadjoint operators in* H_1, H_2 *respectively, then the tensor product operator*

$$A_1 \otimes A_2 : D(A_1) \otimes D(A_2) \subset H_1 \otimes H_2 \longrightarrow H_1 \otimes H_2$$

is well-defined as the selfadjoint operator determined by

$$(A_1 \otimes A_2)(x \otimes y) = A_1x \otimes A_2y \text{ for } x \in D(A_1)\ ,\ y \in D(A_2)\ . \qquad (A.19)$$

Furthermore, if id_i *denotes the identity on* H_i , $i=1,2$, *then*

$A_1 \otimes id_2$, $id_1 \otimes A_2$ *are commuting selfadjoint operators on* $H_1 \otimes H_2$.

The spectral resolution $(E_1(\lambda))_{\lambda\in\mathbb{R}}$, $(E_2(\lambda))_{\lambda\in\mathbb{R}}$ *of* $A_1 \otimes id_2$, $id_1 \otimes A_2$ *respectively, are given by the spectral resolutions* $(\pi_i(\lambda))_{\lambda\in\mathbb{R}}$ *of* A_i , $i=1,2$:

$$E_1(\lambda) = \pi_1(\lambda) \otimes id_2\ ,\ E_2(\lambda) = id_1 \otimes \pi_2(\lambda)\ .$$

The operator $A_1 \otimes A_2$ *may be represented as*

$$A_1 \otimes A_2 = (A_1 \otimes id_2)(id_1 \otimes A_2)\ .$$

([37], p. 243 et passim, [30], p. 298-301, [28], p. 303-304).

The result of Theorem A.21 can be extended without difficulty to any finite numbers of factors. Note that $D(A_1) \otimes D(A_2)$ is a Hilbert space with respect to the corresponding graph inner product.

We conclude our survey with the analogue of Theorem A.14 which is a simple consequence of our previous results.

Theorem A.22: *Let* A_i *be a selfadjoint operator on* H_i , *with corresponding spectral representation* $V_i : H_i \longrightarrow L_i$,

then

$$V = V_1 \otimes \ldots \otimes V_n : H_1 \otimes \ldots \otimes H_n \longrightarrow L_1 \otimes \ldots \otimes L_n\ ,$$

determined as the isometric extension of

$$x_1 \otimes \ldots \otimes x_n \longrightarrow V_1x_1 \otimes \ldots \otimes V_nx_n\ ,$$

is a unitary operator. The functions in $L_1 \otimes \ldots \otimes L_n$ *can be regarded as defined on* $\mathbb{R}^n$; *in this sense we have*

$$V(A_i^{\otimes}\Phi)(\lambda) = \lambda_i\ V(\Phi)(\lambda) \text{ for } \lambda \in \mathbb{R}^n\ ,\ \Phi \in D(A_i^{\otimes})\ ,$$

where

$$A_i^{\otimes} = id_1 \otimes \ldots \otimes id_{i-1} \otimes A_i \otimes id_{i+1} \otimes \ldots \otimes id_n\ ,$$

$A := (A_i^{\otimes})_{i=1,\ldots,n}$ *is an abelian system of selfadjoint operators. For every*

Borel function F *on* $\mathbb{R}^n$ *we have*

$$V(F(A)\Phi(\lambda) = F(\lambda)\ V(\Phi)(\lambda)\ , \quad \lambda \in \mathbb{R}^n ,$$

for $\Phi \in D(F(A))$,

$$D(F(A)) = \{x \in H_1 \otimes \dots \otimes H_n \mid F(\cdot)V(x) \text{ is bounded in } L_1 \otimes \dots \otimes L_n\}.$$

References

[1] Agmon, S.: Lectures on Elliptic Boundary Value Problems.Van Nostrand Mathematical Studies 2, Princeton (1965).

[2] Akhiezer, N.I., Glasman, I.M.: Theory of Linear Operators in Hilbert space. Pitman Publ. Ltd., Vol. 1, London (1981).

[3] Akhiezer, N.I., Glasman, I.M.: Theory of Linear Operators in Hilbert space. Pitman Publ. Ltd., Vol. 2, London (1981).

[4] Antosik, P., Mikusinski, J., Sikorski, R.: Theory of Distributions. The Sequential Approach. Elsevier Scientific Publ. Co., New York (1973)

[5] Beals, R.: Topics in Operator Theory. Chicago Lectures in Mathematics. The University of Chicago Press, Chicago (1971).

[6] Bachman, G., Narici, L.: Functional Analysis. Academic Press, New York (1966).

[7] Bonic, R.A.: Linear Functional Analysis. Gordon and Breach, New York (1969).

[8] Bracewell, R.N.: The Fourier transform and its applications. McGraw-Hill, New York (1965)

[9] Bracewell, R.N.: The Hartley Transform. Oxford University Press, New York (1986).

[10] Dunford, N., Schwartz, J.T.: Linear Operators. Part 1: General Theory. Pure and Applied Mathematics, Interscience Publ., Vol. 7, New York (1963).

[11] Dunford, N., Schwartz, J.T.: Linear Operators. Part 2: Spectral Theory, Selfadjoint Operators in Hilbert space. Pure and Applied Mathematics, Interscience Publ., Vol. 7, New York (1963).

[12] Erdélyi, A., Magnus, W., Oberhettinger, F., Tricomi, F.: Higher transcendental functions, Vol. I. The Bateman Manuscript Project. MacGraw-Hill, New York (1953).

[13] Erdélyi, A., Magnus, W., Oberhettinger, F., Tricomi, F.: Tables of integral transforms, Vol. II. The Bateman Manuscript Project. MacGraw-Hill, New York (1954).

[14] Helgason, S.: The Radon Transform. Progress in Mathematics 5, Birkhäuser Verlag, Boston (1980).

[15] Jörgens, K., Rellich, F.: Eigenwerttheorie gewöhnlicher Differentialgleichungen. Springer Verlag, Heidelberg (1976).

[16] Kasuga, T.: On Sobolev-Friedrichs' generalization of derivatives, Proc. Jap. Acad. 23, 596-599 (1957).

[17] Kato, T.: Perturbation Theory for Linear Operators. Springer Verlag, Berlin (1966).

[18] Koornwinder, T.H.: Jacobi functions and analysis on noncompact semisimple Lie groups, in R. Askey et al. (eds.), Special functions: Group theoretical aspects and applications, I, D. Reidel Publ. Co., Hingham, Massachusetts (1984).

[19] Lebedev, N.N.: Spezielle Funktionen und ihre Anwendungen. Bibliographisches Institut, Mannheim (1973).

[20] Leis, R.: Zur Theorie der Inegraltransformationen. Math. Meth. in the Appl. Sci. 1, 114-125 (1979).

[21] Leis, R.: Initial Boundary Value Problems in Mathematical Physics. Wiley & Sons, Chichester (1986).

[22] Ludwig, D.: The Radon Transform on Euclidean space. Comm. Pure Appl. Math., Vol. XIX, 49-81 (1966).

[23] Magnus, W.,Oberhettinger,F., Soni, R.P.: Formulas and Theorems for the Special Functions of Mathematical Physics. Springer Verlag, 3rd Edition, Berlin (1966).

[24] Mercer, M.: On integral transform pairs arising from differential equations of any even order. Quart. J. of Math. 24, 9-15 (1963).

[25] Olver, F.W.J.: Asymptotics and Special Functions. Academic Press, New York (1974).

[26] Papoulis, A.: The Fourier integral and its applications. McGraw-Hill, New York (1962).

[27] Picard, R.: On some special functions of mathematical physics and their relation to spectral theory. Vorlesungsreihe SFB 72, No. 20, University of Bonn (1984).

[28] Prugovecki, E.: Quantum mechanics in Hilbert space. Pure and Appl. Mathematics, Vol. 41, Academic Press, New York (1971).

[29] Price, J.F.: Uncertainty Principles and Sampling Theorems, in 'Price, J.F. (ed.), Fourier Techniques and Applications', Plenum Press, New York (1985).

[30] Reed, M., Simon, B.: Methods of Modern Mathematical Physics, Vol. 1: Functional Analysis. Academic Press, New York (1972).

[31] Rühs, F.: Zur Theorie der Integraltransformationen. Beiträge zur Analysis 4, 123-121 (1972)

[32] Schäfke, F.W.: Einführung in die Theorie der speziellen Funktionen der mathematischen Physik. Springer Verlag, Berlin (1963).

[33] Szökefalvy-Nagy, B. von : Spektraldarstellungen linearer Transformationen des Hilbertschen Raumes. Ergebnisse der Mathematik und ihrer Grenzgebiete, Vol. 5. Springer Verlag, Berlin (1942).

[34] Titchmarsh, E.C.: Introduction to the Theory of Fourier integrals. Clarendon Press, Oxford (1948).

[35] Titchmarsh, E.C.: Eigenfunction Expansions. Clarendon Press, Oxford (1962).

[36] Triebel, H.: Höhere Analysis. Verlag Harri Deutsch, Thun (1980).

[37] Weidmann, J.: Lineare Operatoren in Hilberträumen. B.G. Teubner Verlag, Stuttgart (1976).

[38] Weir, A.: Lebesgue Integration and Measure. Cambridge University Press, Cambridge (1973).

[39] Wilcox, C.H.: Scattering Theory for the d'Alembert Equation in Exterior Domains. Lecture Notes Nr. 442. Springer Verlag, Berlin (1975).

[40] Yosida, K.: Lectures on Differential and Integral Equations. Pure and Applied Mathematics, Vol. 10. Interscience Publ., New York (1960).

[41] Zemanian, A.H.: Generalized integral transformations. Interscience Publ., New York (1968).

[42] Zemanian, A.H.: Distribution Theory and Transform Analysis. International Series in Pure and Applied Mathematics, MacGraw-Hill, New York (1965).